Marietta Blau – Stars of Disintegration

Biography of a Pioneer of Particle Physics

Studies in Austrian Literature, Culture and Thought

General Editors:

Jorun B. Johns
Richard H. Lawson

Gedruckt mit
Unterstützung des

Printed with support
from the

Rektors der
Universität Wien
Prof. Georg Winckler

Rector of the University
of Vienna
Prof. Georg Winckler

Österreichischen
Bundesministeriums
für Bildung, Wissen-
schaft und Kultur

bm:bwk

Austrian Federal
Ministry for
Education, Science
and Culture

Österreichischen
Bundesministeriums
für auswärtige
Angelegenheiten

Austrian Federal
Ministry for
Foreign Affairs

Brigitte Strohmaier • Robert Rosner

Marietta Blau – Stars of Disintegration

Biography of a Pioneer of Particle Physics

English Edition Edited by Paul F. Dvorak

ARIADNE PRESS
Riverside, California

Translated and expanded from the German
Marietta Blau – Sterne der Zertrümmerung
© 2003 Böhlau Verlag Wien

Library of Congress Cataloging-in-Publication Data

Marietta Blau, Sterne der Zertrümmerung. English
  Marietta Blau, stars of disintegration : biography of a pioneer of
particle physics / Brigitte Strohmaier and Robert Rosner.
    p. cm. -- (Studies in Austrian literature, culture, and thought)
  Includes bibliographical references and indexes.
  ISBN-13: 978-1-57241-147-0
  ISBN-10: 1-57241-147-3
  1. Blau, Marietta, 1894-1970. 2. Nuclear physicists--Austria--
Biography. 3. Women physicists--Austria--Biography. 4. Nuclear
physics--History. I. Strohmaier, Brigitte, 1948- II. Rosner,
Robert W. III. Title. IV. Series.

QC774.B53M3713 2006
539.7092--dc22

2005037944

Cover Design
Art Director: George McGinnis
Photograph courtesy of Eva Connors
Drawing from Sitzungsber. Akad. Wiss. Wien,
Math. Naturwiss. Kl. IIa 146 (1937) 623-641

Copyright ©2006
by Ariadne Press
270 Goins Court
Riverside, CA 92507

All rights reserved.
No part of this publication may be reproduced or transmitted
in any form or by any means without formal permission.
Printed in the United States of America.
ISBN-13: 978-1-57241-147-0. ISBN-10: 1-57241-147-3.
(trade paperback original)

CONTENTS

PREFACE 5

MARIETTA BLAU – THE WOMAN
Childhood and Education 13
First Professional Work – The Radium Institute in the 1920s... 21
Development of the Photographic Method 33
Discovery of Disintegration Stars 44
The End of Marietta Blau's Activities in Vienna 48
Exile in Mexico 54
The Vienna Radium Institute: 1938–1945 67
Move to the United States – Work in Industry 69
Blau's First Research Projects in the United States:
    Columbia University 72
Post-War Austria and the Physics Institutes in Vienna
    after 1945 80
Blau's Scientific Research in Brookhaven and Miami 84
Remaining Years in Vienna 93
Notes: Stefan Meyer, Karl Przibram, Berta Karlik,
          Hans Pettersson, Gerhard Kirsch, Viktor Hess,
          Ellen Gleditsch, Hans Thirring, Georg Stetter,
          Gustav Ortner, Cecil F. Powell, Arnold Perlmutter,
          Chien-Shiung Wu, Herbert Pietschmann 103

MARIETTA BLAU – TEACHER AND FRIEND
Contacts with Marietta Blau in Vienna, 1935–1937
    and after 1960 (Hanne Ellis-Lauda, Vienna) 127
Marietta Blau – My Teacher in Mexico in 1943
    (Pierre Radvanyi, Orsay, France) 128
Collaboration with Marietta Blau at Columbia University
    in 1949 (Martin M. Block, Evanston, Illinois) 130
My Interaction with Marietta Blau at Columbia University
    in 1950 (Seymour J. Lindenbaum, Upton, New York) 130
Memories of a Lengthy Collaboration Beginning in 1956
    (Arnold Perlmutter, Coral Gables, Florida) 132

Recollections of Marietta Blau in Miami, 1960
  (Sylvan C. Bloch, Tampa, Florida) 136
Acquaintance with Marietta Blau in the United States and
  in Vienna (Leopold Halpern, Tallahassee, Florida) 137
Marietta Blau and the Vienna Plate Group
  (Brigitte Buschbeck, Vienna) 138
In Scientific Dialogue with Marietta Blau
  (Herbert Pietschmann, Vienna) 140
Joint Laboratory Work and a Trip to Switzerland with
  Marietta Blau in 1961 (Hannelore Sexl, Vienna) 141
My Doctoral Thesis with Marietta Blau in Vienna:
  1960–1964 (Gerda Petkov, Vienna) 143

MARIETTA BLAU – THE SCIENTIST
Photography and Early Research on Radioactivity 145
Marietta Blau's Initial Investigations 150
First Photographic Detection of H-Rays 151
Selection and Pre-Treatment of Photographic Emulsions 153
Detection of Neutrons and Spectroscopy of Protons
  from Nuclear Reactions 155
Cosmic Rays: Fast Protons and Spallation Stars 157
The *Anschluss* – An Abrupt End to Marietta Blau's
  Research in Vienna 160
Marietta Blau's Findings as a Starting Point
  for New Research 161
Interlude in Oslo 162
Blau's Publications in Mexico 163
Move to the United States 168
The First Scintillation Counter 168
Research on Radioactivity 169
Blau's Return to the Photographic Method in 1948 172
Marietta Blau at Brookhaven National Laboratory 178
Marietta Blau's Research at the University of Miami 181
Review Articles and Other Writings by Marietta Blau 183

BIBLIOGRAPHIES
Marietta Blau's Scientific Publications
 in Chronological Order 189
Selected Literature 196
Internet Sources 204

Index of Names 205
Index of Subjects ....: 212
Index of Publications by Marietta Blau 219

Preface

Marietta Blau was an Austrian physicist responsible for developing the photographic method for the detection of nuclear particles in emulsion plates in the 1920s. This method proved to be one of the most successful tools in investigating nuclear phenomena just at the time when nuclear science was developing rapidly. Blau's life was shaped by the political events of the twentieth century, "the age of extremes." Her scientific career began at the Institut für Radiumforschung in Vienna, one of the major research centers for nuclear science at the time. The German annexation of Austria forced her to leave her native city. With the help of Albert Einstein, she emigrated to Mexico, where women working in scientific fields were uncommon. It was only in the United States, at the age of fifty-four, that she was able to return to scientific research.

The climax of Marietta Blau's scientific achievements was the observation of the so-called "stars" in photographic plates, which she and her collaborator Hertha Wambacher correctly interpreted as tracks of particles originating from the explosions of atomic nuclei in the photographic emulsions, caused by particles of cosmic rays. After Blau was forced to leave Vienna in 1938, the work she had completed up to that point was taken up by Cecil Powell in England and developed further. Powell's efforts finally led to the discovery of the π-meson in cosmic rays, for which he was subsequently awarded the Nobel Prize in Physics. Although Marietta Blau was nominated for the Nobel Prize three times – the first time in 1950 when Powell received it – she was never recognized by the Nobel Committee for the discovery of disintegration stars. At age sixty-six, she returned to Vienna in failing health.

This biography is based on our book *Marietta Blau – Sterne der Zertrümmerung*, published in German by Böhlau-Verlag, Vienna, in 2003. At the time we first began planning a publication on Marietta Blau, her name had been nearly forgotten among scientists in Austria. Therefore, we decided to address the

Austrian scientific community in particular and included scientific essays by Prof. Thomas Schönfeld, Vienna, and Prof. Arnold Perlmutter, Miami, in which the development of Blau's scientific methods are described in some detail. The book also contains facsimile reproductions of three of Marietta Blau's original publications. Blau is now celebrated as one of the great scientists of the University of Vienna; a hall in the university building has been named in her honor.

In consultation with the editor of Ariadne Press, we have revised the scientific part of the original text for the English edition in order to make it more generally accessible. Nevertheless, in order to gain a full understanding of the significance of Marietta Blau's pioneering work in nuclear and particle physics, some information on her scientific approach is essential.

Moreover, we have enhanced the biographical part: many more excerpts of Marietta Blau's letters have been included in order to present a clearer understanding of her personality. We discuss in greater detail the political developments in Austria which shaped Marietta Blau's life and also the circumstances of her life in Mexico.

In view of the growing interest in women scientists and the fact that Marietta Blau contributed to the development of nuclear science at some of the major scientific institutions in the United States, we hope that this book will draw the attention of American readers to this outstanding individual.

ACKNOWLEDGMENT

Many persons contributed to our research on Marietta Blau's life and to the publication of *Marietta Blau – Sterne der Zertrümmerung*, on which the present book is based: Wolfgang Breunlich, Alfred Chalupka, Margarethe Heinrich, Edmund Hlawka, Wolfgang Kerber, Walter Kutschera, Ingrid Meder, Margarete Meschkan (d. 2004), Herbert Pietschmann, Wolfgang Reiter, Anny Schlemko-Frantz, Otto Schwerer and Walter Thirring (Vienna), Gerhard Oberkofler (Innsbruck), Robert Schulmann (Boston), Richard Toeman (d. 2005; London), Joseph Aschner (New York City), Paula Bizberg, Renate Hanffstengel,

Franzi Loewe and Esperanza Verduzco Ríos (Mexico City), and Neva Šlibar (Ljubljana).

We thank Eva Connors, Arnold Perlmutter, Herbert Pietschmann, and Pierre Radvanyi, as well as Agnes Rodhe and Hannelore Sexl for photographs from their personal collections. Further photographs come from the archives of the Österreichische Zentralbibliothek für Physik and those of the Österreichische Akademie der Wissenschaften in Vienna, which are gratefully acknowledged.

We are indebted to Brigitte Buschbeck, Hanne Ellis-Lauda, Leopold Halpern (d. 2006), Gerda Petkov, and Lore Sexl, who were interviewed by Reinhard Schlögl, Vienna, for sharing their memories of Marietta Blau. We also thank Martin M. Bloch, Tampa, Florida; Sylvan C. Block, Evanston, Illinois; Seymour J. Lindenbaum, Upton, N.Y.; Arnold Perlmutter, Miami, Florida; Herbert Pietschmann, Vienna; and Pierre Radvanyi, Paris, who wrote down their personal recollections for us.

In our research we were assisted by the archives of the Vienna Jewish Community, the Rahlgasse High School, the University of Vienna, the Stadt- und Landesarchiv Wien, the Österreichisches Staatsarchiv (Vienna), the Max-Planck-Gesellschaft (Berlin-Dahlem), the Berlin-Brandenburgische Akademie der Wissenschaften (Berlin), the Dokumentationsarchiv des Österreichischen Widerstands (Vienna), and by the Albert Einstein Archive of the Hebrew University (Jerusalem), the Center for History of Science at the Royal Swedish Academy of Sciences (Stockholm), the Kommission für Geschichte der Naturwissenschaften, Mathematik und Medizin der Österreichischen Akademie der Wissenschaften, the Archiv der Österreichischen Akademie der Wissenschaften, the Lainz Hospital, and the Cultural Department of the Mexican Embassy (Vienna).

We are grateful to Alfred Chalupka and Hedy Feierl for their patience and consideration in their repeated reading of the English manuscript at various stages of its formation, as well as for suggesting corrections and improvements.

We are greatly indebted to Jorun Johns for accepting and processing the manuscript with Ariadne Press, to Paul F. Dvorak for his subtle and thoughtful revision of the text, and to Barbara Gable for copyediting the manuscript.

COMMENTS ON THE FORMAT OF THE BOOK

Notes as well as footnotes, i.e., references to journals and books and quotation of other sources of information, are placed at the end of each article. References and quotation of sources are indicated in the text by numerical superscripts, notes by numerical superscripts preceded by an A, e.g., [A1]. Marietta Blau's writings are cited by numerical supercripts preceded by a B, e.g., [B1]; these citations refer to the list of publications (p. 189), which is presented in the bibliographies together with selected literature. In the references and bibliographies, the abbreviation of journal titles conforms to ISO4-1984(E) as far as possible; a few additional abbreviations are explicitly defined.

Indented blocks of text marked with diamonds give either background information or additional information.

Contributions originally written in German or French were translated by Hanne Ellis-Lauda (her personal recollections), Brigitte Strohmaier (foreword by Walter Kutschera, the biographical article by Brigitte Strohmaier and Robert Rosner, the personal recollections of Pierre Radvanyi, Leopold Halpern, Brigitte Buschbeck, Herbert Pietschmann, Hannelore Sexl, and Gerda Petkov). The personal recollections of Sylvan C. Bloch, Martin M. Block, Seymour J. Lindenbaum, and Arnold Perlmutter were originally written in English and are reproduced in the original versions. Letters and other texts in quotations which were originally written in German were translated by Robert Rosner and Brigitte Strohmaier. The chapter on Blau's scientific work relies partly on articles written by Thomas Schönfeld and Arnold Perlmutter for the German edition, the latter having been first written in English.

Vienna, September 2005   Brigitte Strohmaier and Robert Rosner

FOREWORD TO THE GERMAN EDITION
MARIETTA BLAU – STERNE DER ZERTRÜMMERUNG

Biographies bear subjective witness to life. On the one hand, the subjectivity of the person who is portrayed should be depicted accurately. Nevertheless, this expectation conflicts with a taboo in the case of scientists' biographies well into the twentieth century: Scientists were to be known to the public only through their work and their individual lives and personalities were to be kept hidden. At the same time, the description of a person, the inquiries into his or her life, and the interpretation of his or her curriculum vitae and behavior are shaped by the subjectivity of the biographer. The biographer's understanding of the individual presented will depend on a "consonance of souls." And a biography may also be exciting in that it may point to features which went unnoticed while the individual was still alive. Henry Miller is said to have gone so far as to recommend to the potential writers of his biography: "Make it all up! That's the only way to get it right – make it all up!"[1] Perhaps it is a true sign of a great personality that it can be interpreted in different ways. Persons who are essentially unknown to us may appear in a new light that only reveals their importance from the perspective of our day and time. It is in this way that the biography of Marietta Blau, one of the first generation of women active in research into radioactivity and a pioneer of modern particle physics, should be approached. This biography illuminates the darkness which has covered this outstanding woman so heavily that she has remained largely unknown to date, even among experts. Even the writer of these lines has to admit that when he entered his position as head of the very institute at which Marietta Blau had achieved her most important research results in the period from 1923 to 1938, her name was unfamiliar to him.

How can one explain that a woman who developed a method of detection that proved to be one of the most important of early particle physics, nuclear emulsions, has remained essentially unknown? There is no simple answer to this question, but after reading the present biography, one may gain some insight into why this is so. It is interesting to note that the proportion of women engaged in research into radioactivity in its infancy was extraordinarily high. Many of these women succeeded against all odds and achieved outstanding accomplishments, both in science and society. Several of them, such as Marie Curie, became shining stars in the firmament of science and were aware of the role they were playing. Marietta Blau lived a life which avoided attracting any undue attention and generally tended to stay in seclusion. In this sense, she resembled Lise Meitner. Both of these women were confronted with the same hurdle in their day, namely being both a woman and a Jew. It was their overpowering desire for knowledge that made it possible for them to overcome these obstacles. Both had to leave their places of research at the zenith of their scientific careers because of racial policies. The scientific loss was tremendous, but the consequences for their private lives is something to which too little attention has been paid.

Marietta Blau's scientific achievements are closely related to the detection of ionizing particles in nuclear emulsion plates. These investigations were conducted mainly while she worked at the Institut für Radiumforschung in Vienna between 1923 and 1938. After the first detection of protons (then called H-rays), along with the then well-known α-particles, she and Hertha Wambacher were the first to observe the so-called stars in nuclear emulsion plates, which they correctly interpreted as tracks of particles originating from the explosion of atomic nuclei in the photographic emulsions, caused by particles of cosmic rays. After Blau was forced to leave Vienna in 1938, Cecil Powell took up the work on the photographic detection of particles in England and developed it further. This finally led to the discovery of the π-meson in cosmic rays, for which Powell was awarded the Nobel Prize. Similar to Lise Meitner's with respect to nuclear fission, Marietta Blau's research on the disintegration stars left her empty-handed as far as the Nobel Prize goes.

The title of the present book, *Stars of Disintegration*, refers on the one hand to Marietta Blau's pioneering work, but it has a metaphorical meaning as well. Marietta Blau shared the fate of many scientists who, due to the disintegration of Austrian sovereignty, were scattered throughout the world and became stars in their respective scientific fields far from home. This book, whose authors explore the scientific as well as the private side of Marietta Blau's life, conveys the picture of a woman, who, because of her dedication to science and because of the political circumstances of her time, suffered considerably in her private life. Her life, however, is a shining example of how genuine devotion to science conquers even the most difficult conditions. With Marietta Blau's biography, the authors have made an important contribution not only to the history of physics but also to the development of science in our time. The words Lise Meitner spoke during a talk in Vienna in 1963 encompass the essence of what led Marietta Blau to reach for the stars as well: "I believe all young people think about how they would like their lives to develop. When I did so, I always arrived at the conclusion that life need not be easy; what is important is that it not be empty. And this wish I have been granted."[2]

As the present head of the Institut für Isotopenforschung und Kernphysik der Universität Wien, which was founded in 1910 as the Institut für Radiumforschung of the Academy of Sciences in Vienna, I bow in respect before Marietta Blau, one of the pioneering women of radioactivity. With neverending enthusiasm, she shone light upon basic processes of cosmic rays and their effects, which still today form the basis of our institute's activities.

    Vienna, May 2001   Walter Kutschera

---

[1] Erica Jong, *The Devil at Large: Erica Jong on Henry Miller* (New York: Grove Press, 1993), 26.
[2] L. Meitner, "Looking Back," *Bulletin Atomic Scientists* 20 (1964) 2.

MARIETTA BLAU – THE WOMAN

Marietta Blau's life spans an era of tremendous political and social upheaval, as well as one of enormous technical and scientific achievements. In the year she was born (1894), the last Czar seized power in Russia; in Germany, the League of Women's Associations was founded. In 1894 in Austria, women still had not been granted admission to the university, something they only achieved three years later. By the year of Blau's death in 1970, a woman – Hertha Firnberg – had assumed the position of head of the science department, and the proportion of female university students had grown to thirty-three percent. During Blau's lifetime two world wars were fought, the monarchies in Germany and Austria collapsed, and the October revolution, the Austrian Treaty of State, the erection of the Berlin Wall, the Indochina and the Vietnam War, the student movement in Paris, the entry of Warsaw Pact troops into Prague, as well as the landing on the moon of U.S. astronauts all took place. During that same period, aviation and wireless telegraphy were invented; radioactivity, the Planck constant, and cosmic rays were discovered; the theory of relativity and the atomic bomb were developed; and numerous elementary particles – electron, proton, and neutron, and their antiparticles, both muons, various π-, K-, and ρ-mesons, several neutrinos and others – were discovered and investigated. Within the history of science, it was only in 1987 that interest in Blau began to emerge. Since then, Austrian authors have published three biographical articles on her.[1-3] In the United States, where Blau worked from 1944 until 1960, further treatises dealing with her life and work have been published during the past few years.[4-8]

CHILDHOOD AND EDUCATION

Marietta Blau was born on April 29, 1894, the daughter of the lawyer Dr. Markus Blau (with the imperial-royal Austrian title of "k. k. Hof- und Gerichtsadvokat") and his wife Florentine, née Goldenzweig. Theirs was an upper middle-class Jewish family living in Vienna's second district, the Leopoldstadt. Many Jews had settled in this district, which was the location of the Jewish ghetto in the seventeenth century. This district and the neighboring ones bordering the Franz-Josefs-Kai alongside the Danube Canal included one half of all Viennese Jews in 1880 and one third even in 1910. (At the turn of the century, about 150,000 Jews were living in Vienna, 8.7% of the total population.) The apartments in the side streets were usually smaller and inhabited by poorer families, while the buildings along the main thoroughfares and large squares held roomier apartments for the well-to-do. The house where Marietta Blau was born was only a few blocks from that of Lise Meitner. Sigmund Freud, Arthur Schnitzler, and Carl Djerassi also spent their childhoods in this area.

Blau's house (Schmelzgasse 6) was situated in a side street of one of the district's major thoroughfares, but the family soon moved. In twenty years, they relocated five times. The addresses reveal that the family moved to increasingly more elegant parts of the quarter,[9] reflecting the economic rise of the Blau family. In 1912 Markus Blau had his office and apartment in Vienna's central district, on Morzinplatz, a very prestigous address (which became infamous after 1938 when the Nazis converted a hotel on this square to the Gestapo headquarters). Well-to-do Jews began to purchase houses in Währing (eighteenth district) and Döbling (nineteenth district), the so-called cottage quarter.[10] Marietta's family moved to this part of the city later on as well.[9]

Markus Blau came from Deutschkreutz, a village approximately fifty miles south of Vienna in the province of Burgenland, part of Hungary until 1921. Deutschkreutz was one of the seven villages in which Fürst Paul Esterházy allowed Jews to settle at the beginning of the eighteenth century and which received autonomous status. The Jewish community in Deutschkreutz was a very religious one that strictly observed all the traditional regulations. It gave the village the Hebrew name "Zelem." The rabbinical seminary (Yeshivah) in Zelem attracted students from all over Central Europe.

Many Jewish families who migrated to the rapidly developing capital of the Habsburg Empire in the nineteenth century loosened their ties to Jewish tradition. This was also true for the family of Judah Blau, Marietta's grandfather, when he moved to Vienna. He changed his name to Julius and his son's name from Mayer to Markus.

Florentine Goldenzweig's family had been residents of Vienna for generations. Markus and Florentine were married in 1891, and in the following three years, Florentine Blau gave birth to two sons and a daughter. In 1895 at the age of twenty-six, Florentine, besides taking care of two-year old Otto and one-year old Marietta, had to cope with the death of her first-born son, Fritz, a loss that left its mark upon the surviving children as well. In October 1896, the youngest son, Ludwig, was born.

Etta, as Marietta Blau was called all her life by her family and friends, attended a five-grade school staffed by the training college for teachers at Hegelgasse 12 in Vienna's first district from October 1900 to July 1905. During her childhood, Etta's brothers made fun of her short stature (she would grow up to be just slightly over five feet tall). The opinion that these hurtful comments may have made her self-conscious about her appearance for the rest of her life[11] seems to contradict what photographs of her reveal. Starting with the school year 1905/06, she attended first the introductory grade and then the first three regular grades of the private high school for girls run by the Association for the Extended Education of Women (Verein für erweiterte Frauenbildung). This school, founded in 1892, was a novel venture in the education of women in Austria[12] and the first to make it possible for girls to obtain the general certificate of education (*Matura*). This certificate was (and still is) conferred after a rigorous examination in the most important subjects and entitles a student to enroll at any university without further examinations. In the beginning, however, the girls had to take the exams at the renowned academic high school for boys, the Akademische Gymnasium. The girls' school was located at that time in the first district, at Hegelgasse 19. In the second grade, twenty-two of the fifty pupils were Jewish, like Blau, *Mosaisch* in official terminology. Despite her excellent achievement in school, Marietta does not seem to have been a typical pupil: Her classmate Helene Pallester recounted the anecdote that Marietta once brought several mice to school with her.[13] Why her schooling was interrupted during the school year 1909/10, when she received private tutoring, is unclear. At any rate, a letter she wrote mentions a woman tutor she liked sufficiently well to visit twenty years later on a trip to Göttingen in Breslau. In 1910 the high school for girls of the Verein für erweiterte Frauenbildung became a public school and moved to Rahlgasse 4 (sixth district). There Marietta was again enrolled in the fifth grade of the humanistic branch of the high school after having passed an entrance examination. She attended the four upper grades of the school and obtained the *Matura* with distinction in July 1914. Physics was a subject she had only in the third, seventh and eighth grades. She chose German, Greek, civics, and mathematics as the subject areas for her final examinations.[14] At that time, there were only twenty-five students remaining in her class.

♦ Blau's physics teachers were Franz Matouschek, Karl Bruno, and Auguste Tinus. One of Tinus's former students, Margarete Meschkan, who received her *Matura* at the Rahlgasse High School in 1928, remembers her education there.[15] She describes Tinus as a kind-hearted woman who, however, kept her distance, addressed the pupils in the formal manner, and appeared to be straight-laced, both literally and in the sense of being reserved. Although she did not always succeed in conveying the essence of the subject, she certainly knew how to motivate the students. The fact that numerous graduates of the humanistic high school turned to science studies later on was not to be interpreted as an indication of especially inspiring instruction in Tinus's lessons but that a humanistic education was considered routine, even by those primarily interested in other subjects. Tinus and the other female teachers are said to have taken great interest in their pedagogical task: as pioneers in women's university education themselves, many of them unmarried,[16] they had devoted themselves entirely to the girls' education and dedicated their love in addition to their intellectual strength to passing on to their pupils the knowledge and appreciation for learning that would enable them to enter university. ♦

It was in the fateful year 1914 when Marietta Blau enrolled at the University of Vienna as a regular student. The Blau family, too, was affected by the outbreak of World War I. Etta's brother Otto, who studied law and had fulfilled his regular army service right after his graduation from high school in an officers' training unit (*einjährig Freiwilliger*), was called up immediately after the outbreak of the war. After he was wounded in September 1914, he was sent home to recover, but in 1915 he was back on active duty at the Russian front. He was only able to return home on leave to finish his studies a few months before the end of the war. Etta's brother Ludwig was called up after graduating from the technical college, TGM, in 1915 and joined an artillery unit in which he served right up to the end of the war. Etta's cousin Franz Weinberger, who was exactly the same age as Etta, was killed in action in 1917.

Marietta had chosen physics and mathematics as her subjects, although her "dream had always been to become a child psychiatrist."[17] In the winter semester of 1914/15, twenty-two girls from her high-school class were enrolled at the University of Vienna, nine of them in science (four in physics and mathematics) and five in medicine.[18] The absence of the young men who had been drafted opened up new opportunities at the universities for women who were not kept from studying because they had to contribute to the family income. The proportion of female students rose significantly in the academic year of 1914/15 and kept increasing during the war years, until it fell drastically in 1919/20. During the course of her studies in 1916, Blau suffered from tuberculosis and had to be treated in a sanatorium.

Marietta Blau attended physics classes given by the professors Erich Lecher, Stefan Meyer, and Felix Ehrenhaft and mathematics classes given by Gustav von Escherich, Gustav Kohn, and Philipp Furtwängler. Later on, she took part in a two-semester laboratory course at the Second Physics Institute and a one-semester course at the Institut für Radiumforschung (Institute for Radium Research).

The history of this institute was quite unusual. While nearly all academic institutions in Austria were established and run by the state, the Institute for Radium Research was founded with a private donation in 1908. Physics institutes had existed at the University of Vienna since 1848. At the beginning of the twentieth century, they were located in a rental house in Türkenstraße in the ninth district. Stefan Meyer and Egon von Schweidler had begun investigating radioactivity as early as 1899, only three years after its discovery by Henri Becquerel. Two years later, in 1901, the Imperial Academy of Sciences established a commission to investigate radioactive substances. The work of Meyer and Schweidler provided the incentive for the Academy to extract radium from ten tons of tailings from the uranium production in Joachimsthal in Bohemia (at that time part of the Austro-Hungarian Empire). The Academy therefore owned the largest quantity of radium in the world. Because of the inadequate technical equipment at the old physics institute, the true value of this resource could not be exploited. At that time, Dr. Karl Kupelwieser, a lawyer, donated 500,000 crowns (about two million dollars) for the construction and equipping of a building devoted to the physical investigation of radium,[19] the Institut für Radiumforschung der kaiserlichen Akademie der Wissenschaften (Fig. 1). The Institute was built on a site between Währingerstraße and Waisenhausgasse (later Boltzmanngasse) in the ninth district. It was dedicated in October 1910 and was called Radiuminstitut for short from the beginning.[20] According to the terms of the donation, the staff of the Radium Institute was employed partly by the Academy of Sciences and partly by the university.[19] In 1912 the new physics institutes of the University of Vienna were built adjacent to the Radium Institute, enabling close cooperation between the Radium Institute of the Academy of Sciences and the physics institutes of the University of Vienna. A few years later, the chemistry institutes were built on the other side of the Radium Institute.

It was at the Second Physics Institute, directed by Franz Serafin Exner, that Marietta Blau worked on her doctoral thesis on a radiological topic, the absorption of divergent γ-rays. The thesis was approved by Franz Exner and Stefan Meyer in July 1918 after it had been accepted for publication in the proceedings (*Sitzungsberichte*) of the Imperial Academy of Sciences. This work,[B1] in which Blau explained the discrepancy between the existing data in the literature on the absorption conditions of bunches of divergent γ-rays by the production of a high-energy secondary radiation unknown up to that time, was the prelude to a long scientific career during which she continued to solve experimental, methodological, and also theoretical problems with the same thoroughness.

Blau passed her final examinations in March 1919 with distinction (unanimous); her Ph.D. graduation took place on

Fig. 1: The newly built Radium Institute in 1910, before the physics and chemistry institutes were built. The view is from the rear courtyard (Österreichische Akademie der Wissenschaften, Archive).

March 29, 1919.[21] Subsequently, she worked on theoretical problems of radioactive radiation and then spent several months as an observer with Guido Holzknecht at the Central X-ray Institute of the General Hospital in Vienna.

At the time Blau was finishing her studies in late 1918 and early 1919, far-reaching changes were taking place in Europe. At the end of World War I, the Austro-Hungarian Empire disintegrated and in its place the states of Austria, Czechoslovakia, Poland, and Hungary emerged. Vienna, formerly the capital of an empire with fifty-three million inhabitants, was now the capital of a small, impoverished country with a population of six and a half million. The infancy of the newly constituted Austrian Republic was accompanied by widespread starvation and considerable political unrest.

At the end of 1919, Markus Blau passed away after a "brief, serious illness"[22] at the age of sixty-four. Besides practicing his profession as a lawyer, he had also published pieces of music. Among the Jewish middle class of the fin-de-siècle and also during Marietta's entire life music played an important role. Markus Blau's brother-in-law, Josef Weinberger, had founded a publishing house for music and co-edited Gustav Mahler's works. As Josef Weinberger's only son had been killed and his two daughters were disabled, Etta's brother Otto, a lawyer like his father, joined the publishers and became manager soon thereafter. He served as director until the company was "Aryanized" in 1938. After World War II, he returned to his position with the company.

Marietta's brother Ludwig withdrew from the Jewish congregation in 1922 and legally declared himself an atheist, unaffiliated with any religious denomination.[23] At the turn of the twentieth century and after World War I, anti-Semitism played a significant role in Vienna. Having lost all connection to their religion and even considering it a hindrance to their social aspirations, many Jews abandoned the Jewish community. Some even changed their family names. Thus, the Goldenzweig family assumed the name of Golwig.[24] Two brothers of Florentine Blau were baptized before 1918 and in this way received new first names.

First Professional Work – The Radium Institute in the 1920s

In the second half of 1921, Blau was employed as a physicist at the x-ray tube factory of Fürstenau, Eppens & Co. in Berlin, where she conducted investigations in electrical engineering and spectral analysis.

She reported to Meyer: "Above all, I respectfully wish you a pleasant vacation [in Bad Ischl]. I have adjusted quite well to Berlin and can even make both ends meet, as far as my salary goes. Also the work has been very agreeable up to now; I mainly deal with x-rays and ultraviolet light. However, I am quite exhausted because I have to work in the laboratory from 8 a.m. to 5:30 p.m. with only half an hour for lunch."[25]

Blau gave notice at the firm in Berlin when she obtained a position as assistant professor at the Institute for the Physical Bases of Medicine at the University of Frankfurt/Main. She started there on January 1, 1922, and was in charge of scientific investigations for the electrotechnical and electromedical industry, as well as conducting theoretical and practical x-ray training for doctors and instructing doctoral students. Together with Kamillo Altenburger, she published papers on the absorption[B2] and theory of the effect[B3] of x-rays. When her mother fell ill in the autumn of 1923, she resigned her position and returned to Vienna. Professionally, she exchanged a paid job for unpaid work at the Institute for Radium Research and at the Second Physics Institute. As a result, she had to rely on financial support from her family since taking care of her mother became her first priority and was to be a determining factor in her life over many decades.

The Vienna Radium Institute was one of the most renowned institutes worldwide for research on radioactivity. Only a couple of years before Blau started working at this institute, Viktor Hess and Georg von Hevesy, both of whom received the Nobel Prize later on, had conducted research there, and Otto Hönigschmid had determined the atomic weight of radium. Despite all the difficulties the Institute had to face in an impoverished Austria, it nevertheless attracted scientists from within Austria and abroad.

Vienna had recovered from the worst of its plight by the time Marietta returned to her home town in 1923. However, the conflict between Austria's conservative government and Vienna's social-democratic city council had become a dominant issue. On several occasions, the conflict between conservative groups and social democrats led to armed clashes.

In Vienna, the city council introduced new taxes to finance the construction of modern apartments for working-class families on a large scale and froze rents for existing apartments. These measures affected the many middle-class citizens who had invested their savings in real estate. This segment of society had already suffered losses due to the collapse of the empire and to inflation. As a result of all these developments, many members of the middle class, particularly students, joined radical conservative and nationalistic organizations. Most of them were anti-Semitic so that Jews tended to support the social democrats, even though they themselves belonged mostly to the middle class and were also affected by the new taxes.

In spite of all these difficulties, cultural life in Vienna flourished. Many cultural activities were subsidized by the city council in order to make classical culture accessible for the working class.

The period from 1923 to 1938, when Blau worked at the Radium Institute, was, however, an extremely difficult one for the sciences in Austria. Members of conservative and German-nationalistic groups held all the critical state and economic positions after 1918 and were not interested in scientific research. Some professorships were dissolved or simply left unfilled. It became increasingly difficult for young scientists to find adequate employment in Austria. Having physicists work on a volunteer basis was customary at the Radium Institute. As of 1914, the Institute for Radium Research reported annually in the almanac of the Viennese Academy of Sciences who worked there, in addition to the persons holding Academy positions (*Funktionäre*). Between 1914 and 1938, 172 scientists are named: some came from research institutes abroad; others were scientific assistants paid by the Ministry of Education or by donations; the majority, however, were doctoral students or graduates who worked at the Radium Institute without pay.

Neither gender nor social class was a criterion, but adequate financial support from their families was required. On a positive note, this custom provided the opportunity, particularly for women, to gain a foothold in physics research. At the Vienna Radium Institute, the percentage of women was especially high (over one third). This has been attributed to the "magic spell" of radioactivity and to the success of Marie Curie at the Paris Radium Institute, but a more plausible explanation is the fact that with the discovery of radioactivity a new field of physics opened up just at the time when women – thanks to their recent admission to science studies at universities – could for the first time acquire the necessary qualifications for physics research and hence have access to a scientific field not yet in the hands of the male establishment. At the Vienna Radium Institute between 1910 and 1938, in particular, women may also have been attracted by the way Stefan Meyer[A1, 26] (Fig. 2) ran the Institute.

For Meyer, co-workers were allies in the quest for knowledge, rather than subordinates, and as a result, his intense personal involvement with them and their needs proved to be a positive influence on the scientific work at the Institute. The gratitude repeatedly expressed towards him in the Institute's numerous publications was well founded. Meyer, born in 1872, was a father figure to many of his collaborators and fostered a family-like relationship among the staff at the Radium Institute. At the Institute's fiftieth anniversary Karl Przibram[A2] (Fig. 3) characterized Meyer's role as follows: "Stefan Meyer's provident wariness, his expertise, his wealth of ideas ... made the Institute what it is; his urbane character smoothed the way for the harmonious collaboration of all involved, and his overflowing kindness of heart clearly tied a bond of friendship around all co-workers; he was indeed the soul of the Institute, which was completely imbued with his spirit."[27]

The "bond of friendship" led his collaborators not only to take an interest in each other's scientific endeavors and to support each other by discussing problems and giving active help but also to share their non-scientific talents with each other. For instance, Berta Karlik[A3] painted watercolors and wrote poems in honor of her colleagues, while Karl Przibram drew caricatures and cut silhouettes of the employees at the physics institutes.

Fig. 2: Stefan Meyer   Fig. 3: Karl Przibram
(Österr. Akademie der Wissenschaften, Archive, Portrait collection).

The cordial atmosphere characteristic of the Radium Institute at that time stood in contrast to the situation in many other parts of the university. The long tradition of anti-Semitism at Austrian universities gained considerable influence in the academic world after World War I. Throughout the 1920s, there were numerous instances of anti-Semitic students assaulting Jewish professors and forcing Jewish students to leave lecture halls.

Not only the German nationalist student organizations (the predecessors of the Nazi organizations), but also Catholic student organizations demanded that restrictions be imposed on the admission of Jewish students.

♦ The representative of the Catholic Student Union, Engelbert Dollfuss, who later became the Chancellor of Austria (and who was murdered by the Nazis) wrote in an article in 1920:

The Austrian universities, particularly in Vienna, should be the centers of Christian South-German culture ... This is the spirit that should come from our teachers. Therefore, it is the holy duty of those responsible to give us teachers from whom we can expect this spirit. Neither a baptismal certificate nor the proper use of the German language guarantees that this is the case. To meet this standard, one requires deep roots in the character of our people, in our culture, and this with one's whole soul ... It is an appropriate student demand that places at the university be reserved first for the youth of our country; we certainly do not want to hinder sons of other peoples who want to be exposed to German knowledge and science at its source from coming here. To the contrary, we offer them German hospitality. They have to demonstrate that they are worthy of it. But we want to preserve our rights as masters of the house and do not want to be pushed aside by one foreign people, one foreign race. While we adhere strictly to the principles of German hospitality, we have to protect ourselves and bar all those parasites who are intruding upon us in proper or improper ways and are beginning to outnumber us. And no dilettant-like solutions will help. All Jews from the East have to be removed and those who are responsible for this situation, the so-called local Jews, restricted to their rights according to the Peace Treaty and in proportion to their numbers. Only in this way can we secure the academic world for our youth and offer hospitality to the youth of many peoples.[28]

This article was not published by a right-wing radical but by the representative of a mainstream student organization in a newspaper of the governing party. ♦

It is not surprising that many Jewish students and university staff, among them Marietta Blau, no longer felt secure when ideas of this kind were becoming more and more prevalent. Blau herself was described in 1926 as "a little Jewess, always looking out for the next pogrom."[29]

In 1924, Blau published a paper on the decay constant of the polonium isotope of mass 218, a nucleus of the radium progeny named RaA.[B4]

♦ Polonium is the element with a nuclear charge number of eighty-four. Each element is characterized by a certain nuclear charge (proton number). Nuclides of a given element, which differ with respect to atomic weight (mass), are called isotopes. All nuclides with a nuclear charge greater than eighty-three are unstable, i.e., they undergo radioactive decay (or spontaneous fission). Some radioactive isotopes emit α-rays, others, β-rays and/or γ-rays. The half-life is the time within which – on statistical average – the number of radioactive nuclei has reached half its initial value and decreases in size the farther removed this isotope is from stability. The same information is contained in the decay constant, a quantity which is in inverse proportion to the half-life. ♦

At this time, Hans Pettersson[A4] (Fig. 4) from Göteborg was already at the Vienna Radium Institute. He had first come in 1922 in order to measure the radioactivity of samples of deep-sea mud and returned to the Vienna Radium Institute every year until 1930. His involvement in marine studies, as well as the importance of polonium in radioactivity research, are reflected in Karlik's painting for the letter "P" (Fig. 5) in her alphabetical illustration of physics topics. "Pettersson, whose dynamic personality brought new verve to the Institute's work, turned his attention – together with G. Kirsch[A5] – to the work on atomic disintegration initiated by Rutherford in 1919"[30] (now referred to as artificial nuclear reactions).

Fig. 4: Hans Pettersson (Österreichische Akademie der Wissenschaften, Archive, Portrait collection).

Fig. 5: Water-color painting and rhyme by Berta Karlik: "Pettersson no doubt travels the ocean. Polonium is sought with devotion." (Archive R. and L. Sexl)

♦ Rutherford had observed that, when energetic α-particles pass through nitrogen, some of the nitrogen is converted into oxygen, and fast, long-range protons are detected. (A proton is a hydrogen nucleus with one positive charge.) These protons were shown to result from the collision of the α-particles with nitrogen nuclei. Subsequent work by Rutherford and Chadwick (1921–1922) determined that many other elements underwent similar transmutations when bombarded with α-rays. As a result of these discoveries, the study of artificial nuclear reactions became the main focus of nuclear research. ♦

At that time, particles emitted in nuclear reactions could be detected only by the scintillation method. This technique relies on the phenomenon that ionizing radiation causes small light flashes on sphalerite (ZnS) screens, which could be observed and counted with moderately powered microscopes. This type of observation required that the eyes adapt to the darkness of the laboratory in order to detect the weak luminosity. The procedure was fatiguing for the observers and hence subject to errors. Therefore, Pettersson and Kirsch soon considered it desirable to improve the methods for detecting fast protons. Pettersson energized the Institute's scientists, not only to develop alternative detection methods to replace the scintillation method, but also to investigate thoroughly the scintillation process itself in order to try to overcome the deficiencies mentioned.

The freedom to become involved in a field of investigation to one's liking, as well as to motivate one's fellow researchers to pursue related tasks, was commonplace at the Radium Institute. Pettersson is quoted as having said: "Here you can do anything."[31]

He suggested to Marietta Blau that she employ the photographic method for detection of the reaction products of these nuclear reactions.

◆ Photographic processes are based on the fact that silver bromide decomposes under the influence of light or of radiation with higher energy (e.g., x-rays, rays from radioactive decay). In practice, emulsions containing small silver-bromide crystals in gelatin are spread on a glass plate (or a nitrocellulose film) as a carrier. The decomposed silver-bromide crystals, whose number is dependent upon the intensity of the radiation, form an invisible latent image in the emulsion. It can be made visible by development which results in elemental silver remaining in the gelatin. The sensitivity of the emulsion can be modified by varying the size of the silver-bromide crystals, by changing the thickness of the emulsion, by adding chemicals which influence the sensitivity, etc. ◆

This detection method had previously been under investigation at the Radium Institute: Wilhelm Michl had explained the effect of single α-particles incident on a photographic plate under a grazing angle in 1914 and had done preparatory work to gain insight into processes in the uppermost layers of the photographic plates.[32] These questions had not been pursued further after Michl was wounded in World War I and died of his injuries in 1914.

The development of the photographic method became particularly important for the research at the Radium Institute when it was discovered that the observation of disintegration fragments with the scintillation method had yielded results that differed from those obtained by Ernest Rutherford and James Chadwick in Cambridge. These discrepancies may have been caused by the fact that the counting was biased by the desire to register as many events as possible and were so serious that Rutherford's collaborator, Chadwick, came to Vienna in December 1927 to check Pettersson and Kirsch's procedure. The Viennese results proved incorrect.[30, 33] It was indeed the fact that detecting particles with the scintillation method was largely influenced by subjective factors that had been the incentive in 1924 to strive for a reproducible and hence objective method of particle detection, such as that made possible by the photographic method.

Marietta Blau (Fig. 6) was particularly close to two of her colleagues, Elisabeth Rona and Berta Karlik.[A3] Rona,[34] who was four years Blau's senior, came from Budapest and had studied physics and chemistry there. After finishing her studies (1911), she had worked in Karlsruhe, Budapest, Berlin, and again in Budapest before Stefan Meyer invited her to do research at the

Fig. 6: Marietta Blau in 1927 (Photo courtesy of Agnes Rodhe, daughter of Hans Pettersson).

Vienna Radium Institute in 1925, when she joined Pettersson's group. During a stay at the Radium Institute in Paris, she had gained experience in making polonium preparations, which she also took charge of in Vienna.

Berta Karlik (Fig. 7), ten years younger than Blau, investigated the dependence of the scintillations on the quality of the zinc sulfide and the nature of the scintillation process in her dissertation at the Vienna Radium Institute. Her main finding was that the integral light from the scintillations is a measure for the ionization occurring in the ZnS crystals. Hence, the visual observation could be replaced by a photoelectric cell and an electrometer. Karlik received her doctorate in 1928 and subsequently her certification in teaching. Immediately thereafter, she completed her probationary year as a high-school teacher.[35] In 1929/30, she started working at the Radium Institute, first on a volunteer basis and from 1933 on as a research assistant.

Fig. 7: Berta Karlik in 1951 (Foto Fayer/Österreichische Akademie der Wissenschaften, Archive, Portrait collection).

Like most of the female physicists at the Radium Institute, including Marietta Blau, Karlik was a member of the Verband der Akademikerinnen Österreichs (VAÖ; Austrian Association of University Women). This association had been founded in 1922 by Dr. Elise Richter,[36] a professor of philology, along the lines of the International Federation of University Women that American and English university women had established in 1919. Richter also served as the first president of the VAÖ. The association was largely supported in its objectives by the Association of Austrian Women's Societies, whose president, Marianne Hainisch,[37] was the founder of the Austrian Women's Movement. The VAÖ joined this umbrella organization in the year it was founded and by 1924 had 205 members.[38]

♦ According to the personal conviction of its founder, the association's goal was not to be antagonistic toward male colleagues but rather to find ways and means to assist and advise women in academia, from which they had been barred for so long.[38] The VAÖ published a journal, counseled female students and recent graduates, and, most of all, assisted them financially during those difficult economic times by arranging for opportunities to provide private tutoring. Additionally, the VAÖ awarded fellowships to members, first through the International Federation of University Women, later through the American Association of University Women. In 1938 the VAÖ was disbanded by the Nazis because it was a branch of an international organization. What remained were many close contacts between university women in Austria and abroad, which facilitated the re-establishment of the association after World War II, principally due to Berta Karlik's efforts. ♦

In 1930/31 Karlik was awarded a fellowship from the VAÖ that enabled her to go to London and Paris to further her education.[38] She used the fellowship to work on x-ray investigations of crystal structures with Sir William Bragg at the Royal Institution of Great Britain in London, but also spent some time at the Cavendish Laboratory in Cambridge directed by Lord Rutherford. In addition, she visited research institutions in Paris, such as the Radium Institute, comprising the Curie and Pasteur institutes, and Louis de Broglie's laboratory. In 1933 Elisabeth Rona and Berta Karlik shared the Haitinger Prize of the Vienna Academy of Sciences: Rona was honored for her methodical studies of polonium and the investigation of the fine-structure constant and of α-rays of actinium progeny, Karlik for her work on luminescence.[39]

Also among Blau's friends was Herta Leng. Even though the latter was working on her doctoral thesis on adsorption experiments with radioelements[40] at the First Physics Institute, she was associated with the Radium Institute through both her research and her personal life.

Although the hazards of radiation and injuries to certain occupational groups were documented rather early, practically no measures were taken to protect workers in laboratories from radioactivity, and researchers handling radioactive matter received high doses and faced high levels of contamination. In the Vienna Radium Institute, the walls of the chemical laboratories were said[41] to have been highly contaminated with radium progeny, namely ^{210}Pb, ^{210}Bi, and ^{210}Po, by 1926, which meant that workers were inhaling radioactive particles. Even though vented hoods were used, a solution of one gram of radium (about hundred thousand times the amount allowed today under similar conditions) for production of ^{210}Pb and ^{210}Bi as source of ^{210}Po stood on a hotplate insufficiently shielded by lead bricks. An instrument room housing Geiger counters for β-measurements and parallel-plate condensers for α-counting was separated from the chemical lab only by a narrow hallway. The instruments detected such a high background level that measurements soon had to be performed in the physics institute in another building.[41]

Stefan Meyer described the progression of injuries caused by α-radiation:

> A typical case occurs during the decanting of emanation-saturated preparations. The small vial is held in the hand, so that the thumb, forefinger, and middle finger of the left hand are affected when pouring strong preparations (200–1000 mg) from one bottle to another. We discovered that, after about one week, erythema, an oozing of blood and the formation of blisters similar to that occurring with burns, appeared. Accompanying symptoms are restricted fingernail growth and the formation of hard callouses on the areas of the skin around the nail. The injuries are very painful, similar to frostbite, and last several weeks to months. For treatment we recommend covering the affected area with aluminum acetate. By no means should the blisters be lanced. Since the skin cannot

regenerate well if the injury is severe, disagreeable scars remain that become partially sclerotic and are accompanied by occasional bleeding. They become painful from time to time, even years later. The musculature atrophies. Recurrences were noticed, even after ten months. Effects on the blood profile were observed in researchers working with strong preparations at the Vienna Radium Institute from 1910 on; the number of erythrocytes increases significantly (leucopenia).[42]

Blau herself had suffered radiation damage to her hands, which caused her pain repeatedly for the rest of her life.[43] Elisabeth Rona[41] mentions Otto Hönigschmid as an example of the danger of lung cancer, which occurs when radioactive material is handled and inhaled, thereby affecting the inner body through ionizing radiation. Hönigschmid, an expert in the field of atomic-weight determinations, produced high-purity radium preparations by fractionation and crystallization, among them standards for calibration of further preparations. Even later on, he worked without protective measures against radiation and did not even wear gloves when demonstrating his working methods. Decades later, he developed lung cancer and lost an entire lung.[41] He committed suicide in October 1945.[44]

DEVELOPMENT OF THE PHOTOGRAPHIC METHOD

Between 1925 and 1932, Blau, in part with female co-workers, published numerous papers[B5, B6, B8–B10, B12, B13, B15–B19] on the photographic effect and its quantification of protons and α-particles, as well as on the processes taking place in emulsions hit by such particles. The main objective was to distinguish clearly between α-particle tracks and proton tracks. She found that the distance between the silver grains was much smaller in α-particle tracks than in proton tracks.

In the 1920s, Blau first met Hertha Wambacher, a manufacturer's daughter nine years her junior (born March 9, 1903). Also having attended the school run by the teachers' training college for women in the first district, she had entered the high school for girls of the Association for Women's Extended Education in 1914, the year in which Blau had left it. Wambacher's curriculum vitae reveals that physics was offered in all grades, even in the humanistic branch, and that many more female teachers were active than when Blau had attended. Wambacher was an excellent student during the first five grades, and although her grades dropped off, she nevertheless completed her *Matura* with distinction in 1922.[14] She started studying chemistry, gave up after two years for health reasons, paused for one semester, and then turned to physics,[45] possibly already influenced by Blau.

Wambacher (Fig. 8) worked on her dissertation under Blau's guidance at the Second Physics Institute beginning in fall 1928. Together they investigated how radiation affects photographic plates treated with certain chemicals before exposure.[46] The impregnation of the plates with these chemicals aimed at reducing the sensitivity of the emulsion to light, β-, and γ-radiation without changing its sensitivity to α- or proton rays. The thesis was approved in December 1930, and Hertha Wambacher graduated in May 1932 after moderate success in her final examinations. Thereafter, Blau and Wambacher worked together for six years on methodical investigations of the photographic method.

Fig. 8: Hertha Wambacher (Archive R. and L. Sexl).

These two women seem to have been complete opposites, both in appearance and character: Marietta Blau, short, delicate, dark-haired, sensitive, and distant, but still purposeful and successful; Hertha Wambacher, tall, strong, blonde, robust, and loud, but less determined and impressive in her performance. Perhaps the fact that they were alumnae of the same school, the Rahlgasse, brought them together just as their different characters attracted and complemented one another.

Particular attention was paid to influencing the sensitivity of emulsions and optimizing the developing process with regard to particles of low-density ionization: "We succeeded, at last, in finding optimum conditions for the emulsions then available and in making the method reproducible and, within certain limits, quantitative."[47] Neutrons, discovered by Chadwick in England in 1932, could also be detected because they transfer energy to protons present in emulsions, which are rich in hydrogen. The protons then leave ionization tracks.

> ♦ In 1932 it was discovered that electrically neutral particles are emitted when the element beryllium is bombarded with α-particles. Together with the positively charged protons, these particles, called neutrons, are the constituents of atomic nuclei. Since neutrons have no electric charge, they exert practically no physical or chemical influence on their environment and can move in any material over long distances. But if a neutron collides with a hydrogen nucleus, it sets it in fast motion. Thus, protons result when a substance containing hydrogen is exposed to a neutron source. Because the photographic emulsion contains hydrogen, the photographic method proved to be particularly suitable for the detection of neutrons. ♦

Due to her explanation of the phenomenon of desensitization of emulsion plates by certain dye materials, Blau was invited to visit Agfa in Germany and, along with Hertha Wambacher, was also awarded the medal of the Photographic Society of Vienna.

In 1929 Blau again spent some time in a sanatorium, Hochegg-Grimmenstein. This did not prevent her from smoking for the rest of her life. Marietta, who "longed for the institute,"[48] was at the sanatorium with Herta Leng, who intended to send "solar results" to Meyer. What was meant were the results of studies of the photographic effectiveness of metals exposed to sunlight.[49]

For the academic year 1932/33, Blau received the International Senior Fellowship of the VAÖ. Starting in October 1932, she continued her research with Professor Pohl[50] in Göttingen, where she conducted studies in crystal physics.[B20] On the way there, she stopped at Leipzig for a visit to Agfa:

> I started my trip not according to my plan, but with a two-week delay because Miss Wambacher and I wanted to analyze at least provisionally the plates that Prof. Eggert[51] had once given us so that I might give a report when I am there. (We found out that these types of plates exhibit the strange property of only yielding rows of dots when desensitized.) ...
>
> I've been in Leipzig since October 2 and first visited Agfa, where I had an appointment. As I had been told in advance, Prof. Eggert was not there. But he had arranged for my visit and I was quite surprised at how amiably and hospitably I was received. One gentleman at Agfa, Dr. Luft, who works on the effect of radioactive and x-rays on the photographic layer, was at my disposal for two days. (In the evening he even showed me around Leipzig.) I was introduced to the various directors and was even able to take a look at the manufacturing rooms, which I would never have dared to hope for. The plant is immensely impressive, e.g., they produce 64 km of film per day, apart from the amateur roll pack and x-ray films. It would take me many, many hours to describe the machines for casting, cutting, and wrapping and to enumerate all the test methods.
>
> They even gave a performance of an amateur color motion picture for me, using new film which produces really good color and costs only 15% more than ordinary black and white film. I was also shown a new reproduction technique for color prints that is not yet on the market, as well as various studies on sound movies. Everything appeared to me wonderfully efficient and ingenious.[52]

Her letters from Göttingen to Stefan Meyer and to Berta Karlik show that within a short time she took part in the seminars and discussions at the institute. There she met some of the well-known German physicists. Among them was the famous theoretical physicist Walter Heitler[53] as well as Eduard G. Steinke,[54] whose remarks on the photographic method were of great importance to Blau's work. Steinke had constructed ionization chambers for the detection of cosmic rays.

The working atmosphere at the Göttingen institute was quite different from that back home:

> This is a dreadfully bureaucratic system here; a different person is in charge of each little trifle. Tomorrow I will – with trembling heart – have my first meeting with Pohl. If you are used to Meyer's friendly greetings in the morning, you cannot help considering Pohl's monarchical nodding rather strange. I would like to get in touch with Franck[55] and Born, but I don't know how I am going to manage.[56]

Blau wrote to Berta Karlik in November 1932:

> Your warning regarding two calling cards came too late because the previous day I had already given one to Pohl. Other than that, I have not gone anywhere yet and met only Heitler, whom I asked to let me attend his classes. Heitler is quite lively and communicative, and curious, too, and so I've talked to him a couple of times. His classes on wave mechanics give me great pleasure, and I have learned a lot indeed. He is a splendid lecturer and presents things very clearly. I like his classes better than Thirring's, which we attended together, and we have gotten further now than we did with Thirring in a whole semester. Thirring's attitude is certainly too classical for this. Please, do not take apart your electron diffraction apparatus, even when you are finished. I would like to investigate the potentials in discolored and colloid crystals with you.
>
> My work here is progressing very slowly. Now my automatic monochromator is causing trouble, and after a delay of more than a week, I don't know for sure yet if everything is working all right again. Also, I'm having

difficulty with different work requiring manual skill, as for instance glass blowing with Supremax glass.

One colloquium has taken place already, at which I met all the celebrities. I was most surprised that Eucken[57] with his thick book is in fact a thin, rather young person. There were talks on disintegration experiments performed with protons in the U.S.A., and on positive electrons. Both projects were received quite skeptically and with a handful of jokes. Then a project on molecule oscillations was presented, whose author made fun of his work himself.

People here speak ill of Smekal,[58] and, in fact, he acts unfairly towards the Göttingen staff. People here seem to try to judge others fairly, and one has the impression that Pohl emanates good will. He is adored by all his people, and his is a completely authoritarian system. I was told that Pohl writes all his students' publications himself, and if you compare them with the same clear style of all the papers published here, this must be the case indeed. They even admire Pohl's financial talent. He does not accept gifts from any firm (e.g., he never took anything from Agfa for the institute) and nevertheless always has the money needed to buy new equipment.[59]

In December 1932, Marietta Blau wrote to Stefan Meyer:

Esteemed Professor:
Since the institute in Göttingen will be closed for Christmas and I dearly miss Vienna, I will be traveling home on December 21. I would like to ask you a favor, esteemed professor. May I take the microscope home for the holidays so that I can continue to work without having to go out for the whole day? I would like to interpret the experiments that I conducted before my departure for Göttingen.

Among other topics, Rabinowitsch[60] reported last week on the papers of Antropoff[61] (noble gas halides). All the men were rather skeptical; only Professor Eucken spoke in support of them. In the discussion, Prof. Franck, Prof. Eucken, and Rabinowitsch conjectured that of all the noble gases the existence of halides is most likely for emanation. The existence of emanation fluoride is said to be particularly plausible. Perhaps Prof. Pettersson and Miss Karlik could look for such a compound during their spectroscopic experiments. Moreover, emanation in its solid state is said to have metallic properties; could this question not be investigated at the Radium Institute?

Thursday I gave a two-hour lecture on nuclear disintegration and neutrons at the colloquium at the Pohl Institute. I tried to show what is being done in Vienna. In a very lively discussion, they said that nobody had known how much had been done in this field in Vienna. I've just come from the large colloquium at which Prof. Pohl requested that I give a brief presentation on photographic questions at H exposures. I used the opportunity to report on the latest work on neutrons by Kirsch and Rieder, which nobody here knew about either. Professor Heitler asked me to get the Viennese papers on nuclear disintegration for him.

My work is progressing slowly, but I have really only been working for one month because I needed four weeks to get adjusted and make preparations.[62]

Two months later, Blau reported:

Recently Professor Eggert wrote me that he is very much interested in pinacryptol-yellow sensitization and that he would like to work on that subject if I don't have the time to do so in Göttingen. I wrote him that this is being worked on in Vienna and I would have to inquire whether it would be possible to collaborate. I also asked him if he would work with me on the disintegration of lead by cosmic rays, if this is not being done in Vienna, because Steinke recommended it as suitable for using the photographic method. It would be useful if Eggert could help, particularly because he would take over the handling of the plates since he has special experience in the sensitization and desensitization procedures. Moreover, Miss Wambacher has too much work, and I cannot do the photographic work while I am in Göttingen. By the time the plates are analyzed, I will be in Vienna.

> Would you kindly let me know if you would agree to this because I wrote Prof. Eggert that I would first have to ask whether the work will be done in Vienna; also, I do not know if he wants to do all of this or if he is only interested in the photographic part.
>
> It was very cold here with temperatures down to −18° C [0° Fahrenheit]; now everybody is complaining because it's raining all the time, but I prefer it that way. Anyway, I had more sun and ultraviolet rays than I wanted because I had to work all day with a quartz lamp.
>
> We are insulated from the political situation here because we are not allowed to talk politics at the institute.[63]

Meyer replied[64] that the members of the Radium Institute had discussed Blau's letter and they thought there was no reason why the photographic work should be done anywhere else, particularly since Wambacher had already spent so much time on the topic.

Modestly, Blau wrote in a final letter from Göttingen that she had conveyed to Eggert what Meyer had told her, continuing: "I am so happy that the experiments will be continued in Vienna, but I was not sure whether this would be possible. When I talked with Miss Wambacher about the photographic experiments at Christmas, I was told that possibly there would be no time for them because everything is now concentrated on neutrons."[65]

Following her time in Göttingen, Blau began working in April 1933 with Marie Curie at the Paris Radium Institute. There she continued her investigation of neutron emission induced by α-particles in beryllium,[B22] which she had begun in Vienna in 1932.[B18, B19]

In her letters to Stefan Meyer from Paris, Blau gave a vivid description of her life there, of her work at the Curie Institute, and of the people she met, particularly Marie Curie, her daughter Irène, and Frédéric Joliot.

> I was very warmly welcomed by Madame Curie, who told me that I should propose a theme on which I would like to work but that she would prefer that I do some work with the photographic method. Irène Curie is also very nice but not as

nice as Madame Curie or Monsieur Joliot. I wanted to work on the relationship between range and velocity, or stopping power, of fast H particles from neutron scattering. But there is some disagreement about that with Irène, who proposed projects that are now being worked on in Vienna and which I refused to do ...

Joliot has shown me some interesting equipment but not the Wilson apparatus as of yet ... Rosenblum,[66] whose equipment in Bellevue I have not yet seen, told me that he tries to analyze spectral lines with a counter. The application of Geiger-Müller counters, etc., is said to have been very difficult to introduce here because Madame Curie did not trust them and has only now allowed them to be used. I am invited to take a look at de Broglie's institute on Monday.

Today was a session of the International Federation of University Women to which I was invited. There were many women here from England who spoke with great affection of Dr. Karlik.

The only unpleasant side of Paris is that one is not allowed to smoke in the Curie Institute. The only one who does smoke is Joliot, but when his wife sees him doing so, she gets very annoyed.[67]

Despite her previous plans, Blau decided not to return to Pohl's institute in late summer of 1933 because of the political situation. Hitler had become Reich chancellor in January 1933, and, recognizing the dangers, she went directly to Vienna. She wrote to Meyer:

Today I made up my mind at last and wrote Prof. Pohl that it would be difficult for me to return to Germany now. I wavered back and forth a lot, and I am very sorry because of the work, but recently I've talked to people who came from there and who had news and spoke of the difficulties. I hope you, esteemed professor, will not reproach me for my decision, and also not Prof. Przibram,[68] who recommended that I go there ...

My work here progresses very slowly because I am still experiencing contamination and the plates are clouded, possibly due to the high temperatures we have had for a few days ... Recently I had to give a lecture on photographic methods. It was very difficult for me, but I found some nice vocabulary in the dictionary and had someone correct my manuscript. And so it went quite well. It was only during the discussion when I had to ad lib that there were two occasions when everybody laughed. On one occasion, even Madame Curie laughed out loud. Towards the end of June, I would like to travel to Cambridge.[69]

Blau intended to continue working in Paris over the summer, but the institute closed. At least, Pohl replied in a friendly tone, as she wrote to Meyer:

I have received a nice letter from Pohl, in which he writes that he understands my point of view completely and that he will try to find someone to complete my Göttingen measurements. A paper was just published here in Paris in Fabry's journal which shows a shape of the spectrum of silver colloids in the UV similar to the one I found in Göttingen with alkali-halide crystals with added silver ...

I asked Madame Curie if she would be in Paris in August. She said this is usually not the case, but she had made no plans for the summer yet, and you, esteemed professor, should kindly let her know when you will come to Paris or France so that a meeting might be arranged.

I have heard from Dr. Rona that Agathe [Stefan Meyer's daughter] will come to Paris at the beginning of August. I'm looking forward to seeing her and hope I can show her some of the sights.

I received a note from Prof. Chadwick saying that I can come to Cambridge between July 10 and August 15, and I wrote to him that I will arrive at the end of July or the beginning of August. Could you not let Agathe come with me to England? I would take good care of her. I will stay about three days in London and three or four days in Cambridge. Both of us could stay in Crosby Hall ... and perhaps Agathe can find a young girl with whom she can go to museums while I visit various institutes.

In case I can work in Paris in August and September, which is not certain, I would take no vacation and would come back to Vienna in October. Unfortunately, it is doubtful that I can get my work ready for publication.[70]

In her last letter from Paris in July she wrote:

Unfortunately, I cannot work in the institute during the vacation period ... In August and September nobody works here. I could possibly start again on September 15, but my finances would only last until the end of September, so I will return to Vienna on August 20. I would like to ask you, esteemed professor, for permission to use the microscope of the Radium Institute in order to analyze my plates from Paris.[71]

Looking at Marietta Blau's research trips, one is struck by how self-confidently she interacted with internationally acknowledged scientists, whereas in her private life, she hardly dared to lay claim to the attention of others. Evidently, the fear of not being accepted, which caused her to shy away from people she did not know, did not exist in the field of physics. There she was an expert, accepted by her colleagues.

♦ Until the end of the nineteenth century identity was defined by status and expressed by posture and style. Various models of conduct were used for self-representation and staging. Since the fin-de-siècle, individuality and spontaneity gained importance, manifest in personal appearance and facial expression. ♦

When Blau returned from Paris in 1933, the political situation in Austria had completely changed. The conservative government under Dollfuss had dissolved parliament and also forbidden several political parties and organizations (including the Nazi NSDAP, the Nationalsozialistische Deutsche Arbeiterpartei). Finally the Social-Democratic Party, the main opposition party, was outlawed in February 1934. The resistance by the social democrats was put down by force, including the bombardment of working-class apartment houses. A new autocratic rule was established, shaped along the lines of Mussolini's fascist state.

In 1935 and 1936, Blau tutored five doctoral students[72-76] at the Radium Institute. "For their study of the photographic effects of α-radiation, protons and neutrons," Blau and Wambacher were awarded the Ignaz L. Lieben Prize of the Vienna Academy of Sciences in 1937.[77]

The Ignaz Lieben Prize had been established through a foundation from the bequest of Ignaz L. Lieben, a Jewish banker who died in 1862. The prize for Austrian physicists, chemists, or physiologists was initially awarded triennially (first in 1865), and then, after a further donation from the Lieben family, annually (starting in 1901). Grants were awarded through the Vienna Academy of Sciences. The prize money was thousand schillings in 1927, which corresponds to about ten thousand dollars today. The prestige attached to the award of the Lieben Prize was considered more important than its financial value. The list of recipients includes four Nobel Prize winners and, among others, the physicists Lise Meitner, Friedrich Paneth, Stefan Meyer and Karl Przibram. In 1862, the Ignaz Lieben Foundation intended the prize to continue in perpetuity; however, it was discontinued in 1938 after Austria's occupation and not reestablished after the war.[78] It was only in 2004 that the prize was reinstated with a new donation to the Austrian Academy of Sciences by the U.S. American chemist and industrialist Alfred Bader, who was forced to leave Austria in 1938 as a child.[79]

Discovery of Disintegration Stars

In Blau's opinion, further development of the photographic method required increased thickness of the emulsion layers in the photographic plates in order to register as large a fraction of ionization tracks of high-energy particles as possible. To this end, Blau directly contacted Ilford Ltd. in England. New developing techniques were needed for the increased layer thickness.

Blau had been trying to detect particles of cosmic rays with photographic emulsions since 1932. In 1937 she and Wambacher turned to Viktor Hess,[A6] the discoverer of cosmic rays, who at that time was professor at the University of Innsbruck and ran an observatory at Hafelekar. On this 2300-meter mountain north of

Innsbruck, the intensity of cosmic rays was constantly recorded. Initially, this observatory was equipped with ionization chambers constructed by Eduard Steinke.

Upon their request to Hess, Blau and Wambacher were allowed to expose photographic plates at the cosmic-ray observatory at Hafelekar. On the plates exposed for several months, a new pattern of tracks was discovered, namely that of several reaction products starting where a cosmic-ray induced nuclear reaction had taken place. Due to the starlike shape of these tracks, they were called disintegration stars (*Zertrümmerungssterne*).

♦ Rudolf Steinmaurer, who worked at Innsbruck University and was in charge of the Hafelekar Observatory, wrote:

Two groups of tracks were found: In addition to long tracks that were attributed to protons of energy not previously observed, short tracks were also found which emanated in a star-shaped manner from a center. The author remembers that Miss Blau, while on summer vacation at a resort in Tyrol in 1937, told him in front of the Innsbruck station of the Hungerburg Railway: "I just received a message from Miss Wambacher from Vienna; she found strange starlike tracks on the plates which had been exposed at Hafelekar, which can certainly not have originated in radioactive contamination." This was the first observation of a disintegration star.[80] ♦

Blau herself reported excitedly to Stefan Meyer:

I apologize for bothering you with institute business during vacation. But we found a disintegration effect (with emission of several heavy corpuscles) by ultraradiation [a synonym for cosmic rays], and Prof. Stetter suggested – after only two specimens had been found – that we should publish the findings in a letter to *Nature* or *Naturwissenschaften*. Meanwhile, Dr. Wambacher, who took a microscope with her to Hainbach, has found a total of twelve such disintegration stars, and I sent a compilation and text to Dr. Karlik asking her to translate it for *Nature*, if she found time, or otherwise to pass it along to *Naturwissenschaften*. Dr. Wambacher is insistent that we should not let others overtake us again, and I hope that you, esteemed professor, will not object.

The disintegration processes look like this: From a single silver grain, several tracks radiate, some of which are quite long (the longest corresponds to 1.76 m in air). We found four stars with three prongs, four with four prongs, and one star each with six, seven, eight, and nine prongs. As an example, I enclose a drawing of a nine-prong star. The bold lines denote tracks rich in grains (only very few have a grain density such as α-tracks have), namely with grain distances up to 0.3 cm (transformed to range in air) (α P [protons from α-induced reactions] have approx. 0.2 cm grain distance), the fine lines, low grain density. The arrows indicate the direction from the emulsion to the glass. The lines are interrupted where they cannot be drawn over their entire length. Regarding single tracks, the longest is approx. 12 m in air.[81]

The values of length data (such as grain distance and range [*Reichweite*], i.e., the total distance traveled by a particle of given initial energy) differ for materials (like air and emulsions) of different density and chemical composition; but they can be uniquely converted from one material to the other. Ranges in air were used to characterize particle energies.

The details of the multiple-disintegration process were still unclear since the theory of nuclear forces was just at the very beginning of its development. The results were published quickly indeed, and immediately, the two scientists turned to Prof. Erich Regener, director of the research station for stratospheric physics in Friedrichshafen on the Lake of Constance, which was run by the Kaiser-Wilhelm-Gesellschaft. Their request to expose photographic plates in balloon flights in the stratosphere was turned down by Regener since one of his co-workers was said to have just conducted similar studies and published preliminary results.

In September 1937 Blau made the same request of Friedrich A. Paneth (1887–1958; Fig. 9), who had held the position of Second Assistant at the Radium Institute for several years before 1920. After holding professorship at the universities in Berlin

Fig. 9: Friedrich Paneth in 1928 (Archive Cornell University).

and Königsberg, he had fled from the Nazis in 1933 and now worked at the Imperial College in London. Blau asked whether she might include packets of photographic plates on his balloon flights into the stratosphere in order that she might register particle tracks of cosmic rays in the emulsions. What followed was a lively correspondence[82] on feasibility, achievable altitude, weight and wrapping of the photographic-plate packets, and optimization of the shape and layer thickness of the emulsions.

In her autobiographical sketch, Blau describes her studies in 1937 and her further plans:

> Encouraged by the experimental results, we planned new experiments immediately. To investigate how the size of stars depends on the atomic number of the disintegrating element, the emulsions were exposed unwrapped and covered by foils of various materials. Also, emulsions were sent to various mountain stations at different altitudes and latitudes. Through Prof. Meyer's assistance, we received a grant from the Academy of Sciences enabling balloon flights with the emulsions. All this was interrupted, however, by the political events in Austria in 1938.[47]

The discovery of the disintegration stars was met with great interest in scientific circles, particularly by the theoretical physicists (see p. 161). Werner Heisenberg had developed a theory about cosmic-ray particles colliding with atomic nuclei leading to emission of a large number of particles. Blau and Wambacher's detection of such events now provided experimental evidence for his ideas; their first publication in *Nature*[B36] furthered the discussion of such processes. The correspondence of the important theoretical physicists Wolfgang Pauli,[83] Markus Fierz, Hans Bethe,[84] and Erich Bagge[85] substantiates their interest in the discovery as well.

Last but not least, Albert Einstein was impressed by Marietta Blau's successful method of studying cosmic rays by photographic means, how she along with Wambacher made nuclear reactions visible in emulsions. In fact, his enthusiasm proved to play a decisive role in Marietta Blau's life, as discussed below.

THE END OF MARIETTA BLAU'S ACTIVITIES IN VIENNA

In spite of the great scientific success the two women had in 1937, the relationship between Blau and Wambacher seems to have deteriorated rapidly as a result of political developments, even before the Nazi invasion of Austria. Ellen Gleditsch[A7] (Fig. 10), a professor of inorganic chemistry at Oslo University who had visited the Vienna Radium Institute in 1937/38 and witnessed the Nazis' attitudes there, worked to rescue Blau from the kind of treatment she was receiving, as is apparent from a letter Gleditsch wrote a year later:

> Dr. Blau is discretion itself and she does not and will not say anything against Dr. Wambacher. But I can tell you that Dr. Blau has been abominably treated by the Nazis and among them Dr. W. It was in fact the difficulties with Dr. W. that in January last year made me ask Dr. Blau to work here for some time; I had heard about them, not from Dr. Blau herself, but from other workers in the laboratory.[86]

On February 14, 1938, four weeks before Austria's annexation, when conditions were still seemingly stable in Vienna, none other than Albert Einstein (Fig. 11) intervened to find a post for

Fig. 10: Ellen Gleditsch in 1929 (Department of Chemistry, University of Oslo).

Fig. 11: Albert Einstein (Archive Österreichische Zentralbibliothek für Physik, Vienna).

Marietta Blau in Mexico. Replying to an invitation to participate in a summer school in Mexico, which he refused for reasons of health and the abstract nature of his studies, he recommended that the money allocated for his visit be used to hire Marietta Blau, who at that time still lived in Vienna and worked at the Radium Institute, as a permanent staff member.[87] He mentioned the modest means her research method required, which made her research particularly suited to a country like Mexico.

But Einstein did not write only this letter. On February 14, 1938, he also wrote to his physician and friend in New York, Gustav Bucky, who was expecting a visit from Gustav Peter, a professor at the Technical University in Mexico. In his letter, Einstein urged Bucky to approach Peter in order to help find a position for Marietta Blau in Mexico. The letter to Bucky reads:

> Hitler wobbles somewhat, but not enough!
> I write with respect to the visit of your friend Dr. Peter from Mexico. Could you ask him to find out whether an

extraordinarily talented woman physicist might be needed, who in Vienna – despite all estimation by experts – is being ousted as a Jew for well-known political reasons. Her subject is experimental radioactivity. The doctoral theses completed there in this field are written mainly under her guidance, although as far as I know, she does not hold a regular position. Her personal data are: Dr. Marietta Blau, born 1894, Institut für Radium-Forschung, Boltzmanngasse 3, Wien 9. She has invented a successful method for studying cosmic rays by photographic means (photographic detection of certain atomic disintegration processes resulting from cosmic rays).

This is not an ordinary case but a really valuable person who would be highly capable of kindling scientific life at any place with modest resources. I do ask you to consider with your friend whether any possibility for Ms. Blau exists in Mexico. Here the competition among the local scientists is probably insuperable, at least in Ladenburg's[88] opinion. Nevertheless, we will try here as well. Please, do not forget![89]

Since Einstein expresses the opinion in his letter that Hitler's position might be weakening, he may have tried to facilitate Blau's employment in Mexico because of the scarce career opportunities for her in Vienna rather than because he expected the Germans to march in. Evidently, he did not anticipate the difficulties she would encounter in Mexico. The opinion that Hitler's position might be undermined as a result of the conflict between the Wehrmacht and the NSDAP was shared by many people at that time. This conflict became obvious when, on February 4, 1938, the commanders-in-chief Werner von Fritsch and Werner von Blomberg were dismissed and Hitler took over the supreme command himself.

Blau planned to stay with Ellen Gleditsch in Oslo during the spring and summer of 1938. Blau's trip first brought her to Copenhagen, where she gave a lecture at Bohr's institute, and to Pettersson's place of employment in Göteborg. She described these events in her letter to Rona and Karlik from Copenhagen, although her trip was overshadowed by her experience of seeing German soldiers on their way to invade Austria:

Dear Elisabeth, dear Berta:
Most of all, I wish to thank you for all your kindness.
Yesterday I gave my talk; it was very bad, of course. It was only during the discussion that I recovered somewhat. I had a nervous breakdown from the long, hard trip. Today I went to the institute, and everyone treated me very cordially. I can already control myself much better at this point, only I despair over having left my mother behind, which was completely the wrong thing to do, I'm sure. Frisch showed me around the town and was very nice. I took a lot of his time which he needs urgently because the cyclotron is to be operative on April 5, namely for Bohr's fiftieth birthday, at least for show ...
Please write everything you know, maybe to Prof. Gleditsch or Dr. Føyn. I will write to Prof. Meyer and Prof. Przibram, too, to whom I send the best regards. If I can sleep properly, maybe I will feel better.[90]

On March 21, 1938, Blau wrote from Oslo to Paneth at the Imperial College in London:

I left Vienna on March 12 at seven o'clock in the evening, and I was not really clear about the political situation. I should have left at the beginning of March but postponed my departure several times and perhaps was the last Austrian to pass the German border. In Vienna, we did not know what lay ahead until the last moment, and it was only on my trip that I met the German troops and realized that all hope was gone. I don't know now whether I will ever return or will be treated as a refugee, and I am of course completely desperate. Here in Oslo and in Sweden with Prof. Pettersson and also at the Bohr Institute, where I gave a talk, they were especially kind to me, but I have no idea how things will proceed.
Today I received a letter from Dr. Karlik saying that things have remained the same at the Institute and that I am getting accurate information in the foreign newspapers. I would be very grateful if you could write me when you learn about Prof. Meyer and Prof. Przibram. I don't dare write in order not to create any problems.[82]

Subsequently, Blau and Paneth's correspondence[82] dealt not only with the details of the plates to be exposed in balloon flights but also with the fates of the staff members at the Radium Institute, about which Blau learned from Berta Karlik, albeit in somewhat cryptic form. It was important to assess the political stance and hence the trustworthiness of colleagues transmitting the news but most important of all to plan how to help threatened colleagues find jobs abroad and a way to emigrate. A letter from Blau to Paneth from April 23, 1938, reads:

> I would never have thought that people in Vienna, like Kirsch, Stetter, Schintlmeister (who is now becoming a university lecturer) would behave like this, especially since Prof. Meyer treated them like a father. On the other hand, Prof. Thirring and Franziska Seidl are said to be entirely decent and visit Prof. Meyer, and most of all Berta Karlik, whose character is well known and for whom one has to be afraid because she becomes so personally involved.[82]

In the same letter, she tells Paneth that Meyer was a broken man after his sister's suicide, that his son needed to be able to continue his law studies in England, and that his daughter was not assigned a topic for a dissertation. Later letters deal with contacts with Scandinavian and Swiss physicists and institutions that might possibly invite Viktor Hess and Hans Thirring[A8] (Fig. 12), who had lost their positions because of their opposition to the Nazi regime. In her scientific work at that time, Blau investigated[B41] the α-radiation of a nuclide, which was – in her own words[47] – later on identified as a hitherto unknown samarium isotope. In Germany, shortly after the Nazis had seized power, all Jewish physicists as well as all other Jewish scientists were expelled from the universities. This meant that fifty physicists with *Habilitation* (the right of teaching at universities), i.e., 15.4% of all physicists with *Habilitation* in Germany, were laid off in 1933.[91] In this time of worldwide economic crisis, the budgets for universities and research in general were cut even in the major Western democracies, and hence even renowned and internationally famous physicists who emigrated had difficulty finding appropriate positions.

Fig. 12: Hans Thirring (Archive Österreichische Zentralbibliothek für Physik, Vienna).

In April 1938 when it became clear that Blau would not be able to return from Oslo to Vienna, Einstein, with help from the American Association of University Women and other organizations, made an unsuccessful effort to find a position for her in the United States.

However, Einstein's attempts to procure a professorship in Mexico for Blau brought the desired result. Peter had informed the head of the department of technical studies, Juan de Dios Batiz, of Einstein's letter, and he offered Marietta Blau a position as a professor for advanced students.[92] At that time, Gustav Peter was one of the leading radiologists and cancer specialists in Mexico. He was a Swiss citizen, a learned and gifted man, who in his spare time composed music, painted, and occupied himself with physics.[93] From a letter Marietta Blau wrote to Einstein on June 16, 1938, we learn that she was informed of her appointment at the Technical University in Mexico City by a professor whom she thought to be an Austrian.[94] (Marietta Blau had never achieved *Habilitation*; in Vienna the *venia legendi* [right of teaching at universities] would have been a prerequisite for a professorship.)

The invitation to Einstein to participate in the summer school in 1938 had been the result of the political developments in

Mexico after the election of Lázaro Cárdenas in 1934. Under his presidency, a new awareness, both political and cultural, began. Therefore, at the beginning of 1938, the plan emerged to hold an international summer school at which prominent European intellectuals would – among other topics – demonstrate modern educational methodology. This was also the reason why Einstein was invited.[95] One important step in Cárdenas' reforms was the foundation of a polytechnical institute in Mexico City. These reforms were received with great interest among the progressive intellectuals. This may have been the reason Einstein considered Mexico a country where scientific research was in its infancy and therefore a suitable place for Marietta Blau to work productively.

Mexico was the only country that formally protested before the League of Nations in Geneva in March 1938 against Nazi Germany's annexation of Austria and that never acknowledged the validity of the *Anschluss*. Beginning on March 3, 1937, the Mexican embassy in Paris was the authority responsible for contacts with Austria and remained so after 1938 as Mexico continued to recognize Austria's existence as such during all of World War II.

Blau now had to get her permit of residence in Norway extended until her departure and obtain a German passport (with the stamp "J"[96]) – presumably at the German consulate in Oslo – because her Austrian passport was no longer valid. She also had to get her mother out of Austria, as she had decided to take her along and to care for her in Mexico. She told Paneth[82] in a letter from July 2, 1938, about her nomination as professor at the Technical University in Mexico. In the second half of September, she mentioned a meeting of several days with Berta Karlik in Sweden; at that time, Blau did not yet have an immigration visa for Mexico and was quite concerned about her relatives being detained without explanation in Vienna.[82]

EXILE IN MEXICO

The planned trip to Mexico via London was further delayed because a final decision on her appointment at the Technical University in Mexico City had not been forthcoming and because she had not received her mother's exit permit. In a letter to Paneth,[82] Blau states that, if her mother came to Oslo before the end of September 1938, they would travel from there to London together, spend one or two days (with her uncle Hugo Golwig), and leave for Mexico on October 7, 1938. In fact, Blau left Oslo in the first half of October and was in Mexico at the beginning of November. Presumably, she flew from Oslo via Hamburg to London, while her mother came directly from Vienna to London, and then they crossed the Atlantic together by ship and headed for Mexico. Karlik commented on her departure in a letter to Gleditsch: "I can't help feeling that there is something rather pathetic about this poor little frail figure, so utterly worn out by one blow after another, now crossing the ocean to start a new life in what still seems to me a somewhat exotic country."[97]

◆ During her stopover in Hamburg, Blau was searched and her research notes confiscated. She assumed this action had been initiated by certain Nazi staff members at the Vienna Radium Institute, as she later told Leopold Halpern,[98] an Austrian physicist, who became acquainted with Blau in 1952 in the United States (see p. 137).[4,5] Probably, the search and seizure was just a routine operation since Jewish scientists were generally not allowed to take scientific materials with them when leaving Germany. Leopold Halpern thought that Blau had used an airship for her trip to Mexico. However, the transatlantic airship (*Zeppelin*) traffic had been stopped after the crash of the *Hindenburg* in 1937. ◆

Marietta's brothers also succeeded in emigrating: from 1938 on, Otto lived in England and Switzerland without ever returning to Austria for any extended period of time; Ludwig and his family settled in New York.

After Blau's departure from Oslo, Paneth asked Stefan Meyer to whom he should send the exposed plates from the balloon flights. This was a subtle question inasmuch as analyzing the plates directly involved the copyright on the results. Stefan Meyer considered having the plates evaluated by Blau in Mexico City, by Gleditsch in Oslo, or by Wambacher in Vienna, but he left the final decision to Blau.[99] She accepted Paneth's offer to have the plates developed at his laboratory in England according to her instructions and then to have them sent to Mexico City for further analysis.

On January 1, 1939, Marietta Blau began working as a professor in Mexico City.

> ♦ Mexico City is located on the plateau of central Mexico 2240 m above sea level. It is the capital of the United States of Mexico, consisting of thirty-one states and the Federal District, where the city is located. The Federal District is detached from the state within which it lies (also called Mexico) and comprises an area of 1482 km^2. In 1938, Mexico City had about one and a half million inhabitants. ♦

Exile in Mexico saved Blau from the Holocaust, but she had to cope with working conditions that were highly unsatisfactory. The institution at which she taught was the Escuela Superior de Ingeniería Mecánica y Eléctrica (ESIME) of the Instituto Politécnico Nacional (IPN). Blau was involved with the program for graduate studies. Right from the start, she was disappointed with conditions at the institute. In a letter of January 31, 1939 (dated by mistake January 31, 1938), she wrote to Paneth:

Dear Professor:

Thank you sincerely for your kind letter, and I would also like to congratulate you on your professorship [at Durham University] about which Prof. Gleditsch has already told me.

I am grateful for your suggestion to develop the plates, as they would certainly suffer if sent by mail, and I do not have an office or lab as yet. I was planning to ask you to send them to Norway or to Sweden, where I thought Dr. Rona could be found, so that she could develop them, but unfortunately it seems that she cannot leave Hungary ...

I hope that once the (developed) plates arrive I will at least have a microscope that I can use. It is very difficult here and one requires much patience and energy. The Mexican authorities themselves are very helpful, and it is only because of poor organization that things do not go quite right.

One also feels the resistance of the rather large number of German gentlemen working here. The countryside and the climate are indescribably beautiful.

I was told that they are looking for Dozent Gross from Vienna to offer him a position as professor, but they could not locate him. They have no physical chemist here, and they need one urgently. If you happen to know where he is, please let me know so that I can give his address to the Ministry of Education; possibly he would be very happy about an offer. It's entirely impossible for anyone to come here who is not invited by the government. Particularly in the middle class, there are strong sentiments against immigration.[82]

The "German gentlemen" were to be found even among the professors of the Instituto Politécnico Nacional: several of them were of German origin and at that time were supportive of the Nazis.[100] Regarding the beauty of the scenery and climate, one has to remember that Mexico City had not yet become the smog-filled megalopolis it is today. One could still see the beautiful snow-covered volcanoes Popocatépetl and Iztaccíhuatl from the city. The immigrants from Europe were not only impressed by the rich Mexican culture, but evidently they also enjoyed the pleasant climate.

Initially, Blau had to learn Spanish and acclimate herself to the mentality of the people. Although the ESIME was a technical university, Blau had little opportunity for scientific activities beyond her teaching. Even for her classes, the most basic equipment was missing. In order to do any research at all, she built Geiger counters with the help of support personnel,[92] but the room in which she worked had a leaking roof with glass windows so that measurements could not be made during rain or at noon.[101] More expensive equipment could not even be considered. She used every opportunity to acquire data. For instance, she got her students to take photographic plates for registering cosmic rays with them on mountain trips, occasionally even to Popocatépetl, a peak with an altitude of 5400 m, which is quite close to Mexico City and comparatively easy to reach.

Blau (Fig. 13) had an impressive capacity for work. Denied the chance to conduct scientific research, she turned to the applied

Fig. 13: Marietta Blau 1941 (Identity card, Mexico City, 1941).

study of various phenomena in the part of the world in which she now found herself. For instance, based on theoretical considerations, she estimated the effect of solar radiation on the health of the Mexican people, who are much more highly exposed to the sun than Europeans because a high percentage of the population lives at an altitude above thousand meters in a tropical zone. Because of the considerable seismic activity in Mexico that produces frequent earthquakes and numerous volcanoes (one of which erupted for the first time in February 1943), Blau became interested in questions concerning the earth's crust.

The foundation of the commission for proposing and coordinating scientific and technological research (Comisión Impulsora y Coordinadora de la Investigación Científica, CICIC) gave Blau the opportunity to apply her experience for the advantage of her host country. She was invited to join this organization and headed its radioactivity laboratory. In this position, she studied the radioactivity of minerals and springs in several parts of the country, particularly in the state of Chihuahua, where minerals containing uranium were found in non-working mines. Blau reported on her investigations not only in the Mexican journal *Ciencia* but also in the *Yearbook of the American Philosophical Society*.

In Mexico, Florentine Blau ran the household (which always included a Chihuahua) with the help of a maid who also did the cooking. Except for the days when her adult education classes started early in the afternoon, Blau came home by bus for lunch and returned to the school afterwards. (The house in which mother and daughter lived in 1942 is shown in Fig. 14.)

The relationship between Marietta Blau and her mother seems to have been one of mutual dependency. As mentioned above, Blau had given up her employment in Frankfurt in 1923 to return to Vienna when her mother fell ill, and she was the sibling who emigrated with Florentine Blau and took care of her living expenses. For her part, Florentine Blau (Fig. 15) seems to have provided a certain shelter for her daughter. Blau was said to have had difficulties in coping with everyday life and to have been sheltered through most periods in her life, first by her mother who felt she was different from other people, in later years by her brothers as well.[23] Her mother's presence in Mexico may have eased the confrontation with the practical aspects of life for Marietta. She even hoped to bring Katharina, the younger of her two Weinberger cousins, to Mexico. But this was impossible[97] since she suffered from Down's syndrome and would not

Fig. 14: House (Citlaltepetl 36) in which Marietta Blau lived in 1942 (Photo by Paula Bizberg, 2000).

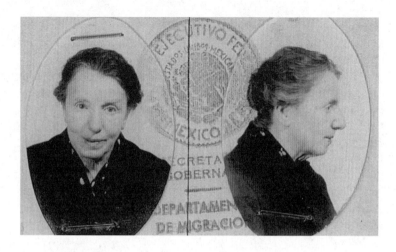

Fig. 15: Florentine Blau in 1939 (Identity card, Mexico City, 1939).

have been able to get a visa. Furthermore, her other cousin, Margarethe, was considered insane. Both cousins perished in a concentration camp in 1941.[102]

Blau's job at the ESIME was hampered by competition and controversy between scientists and engineers.[92] The latter focused solely on industrial progress and had no appreciation for pure science. This group persuaded the director of the institute of their opinion and thereby undermined Blau's position. Moreover, the academic atmosphere was one of male dominance, which was characteristic of Mexico at that time. In a photograph[92] of the teaching staff of ESIME (Fig. 16), Blau is seen on the far left, the only woman among fifty men.

In summer and fall of 1939, some of Blau's acquaintances[103, 104] turned to Albert Einstein again in order to intercede for her because she had not been paid even her minimal salary of approximately hundred dollars per month. These individuals held the opinion that "because of her retiring nature and almost pathological modesty she seems to be incapable of looking after her own interests."[103] Thus once again, others intervened on Blau's behalf more than she herself did: "Since, based on my experiences in Mexico, Dr. Blau will have no other choice than turning

Fig. 16: Teaching staff at the Escuela Superior de Ingeniería Mecánica y Eléctrica of the Polytechnical Institute in Mexico City around 1940 (Archive Keller family).

to you, I am doing the same right now, for Dr. Blau would presumably write only when it is almost too late."[104] In fact, Blau did write to Einstein in 1939 and again two years later.

In 1941, she wrote that she had given a series of lectures at a provincial Mexican university in Morelia and that they had offered her a teaching position there,

> which I would have accepted happily, being the only physicist in Morelia, and far from all competitive bickering from which all foreigners suffer, I would have had more opportunities for work than in Mexico City. What attracted me most was a physics laboratory that the rector – an Indian with lofty ambitions for "his" university – had just purchased and which was sitting there in boxes as there was no one capable of installing it properly. The rector asked me to come to Morelia at least for the time needed to install the apparatus and said the rest would work itself out. Before that, I definitely had to go to Mexico City to settle the matter in the ministry, etc.

> Back in Mexico City, I was offered a permanent teaching position at the polytechnical school, which ensures a secure salary without excitement and bother, but excludes any chance of pursuing my work. What they require is approximately twenty-four hours of teaching per week, more or less on a high-school level. I consulted friends and also my superiors, and they all urged me strongly to accept the post and viewed it as the only safe job I could ever hope for in Mexico. Meanwhile, I got news from Morelia that the apparatus, or at least part of it, had disappeared from the university and reappeared in a pawnshop. The professor responsible went on teaching as if nothing had happened. The salary for the first few months of this year, which I had "made sure of" by running to at least a thousand different places or people, has become uncertain again due to a law that forbids payment of overdue salaries, and how and when I shall receive the "safe payment" from the polytechnical school is somewhat unclear as well. But worst of all: even if things go all right, I will be able to live on the salary together with my mother, but I won't be able to work.[105]

It is clear that by "work" she means scientific research and not the heavy teaching load. In over five years in Mexico (until May 1944), she published seven papers,[B42–B48] the subjects of which were not the photographic method but rather more general topics such as solar radiation, origin and location of helium, measurement of small ionization currents, and the radioactivity of the earth's crust. Later she reported that she had discovered radium deposits that were being mined.[106]

In a letter, Blau commented about the Indians in Mexico:

> Unless they are very poor, they are not dirty at all, and, once they overcome their mistrust of the foreigner and have confidence in him, they are very loyal and quite cheerful. The children are the nicest, and extremely smart ... Many of them are so poor that they have to earn their own living even at the early age of twelve. This is also true for my students. When I reproached one of them for not working harder, he said he had to earn money for his three children.[107]

Besides her professional work, Blau maintained close contact with other immigrants from Central Europe. There were relatively few in Mexico in 1938 when Blau arrived because Mexico was still strictly limiting visas issued to refugees from Germany and Austria. However, a few intellectuals, such as the famous painter Diego Rivera, exerted their influence to help some well-known musicians and artists come to Mexico. The Viennese conductor Ernst Römer, who had been a student of Arnold Schönberg, was welcomed personally by Rivera when he arrived at Santa Cruz. When Blau came to Mexico, she soon became a friend of the Römer family and visited them frequently. In this way, she got to know Diego Rivera, at the time an admirer and friend of Leon Trotsky, who had come to Mexico in 1937.

Many years later Blau recollected that the famous exile Leon Trotsky was in the circle of intellectuals she frequented. In that group as well, was an erratic young man who was later identified as Trotsky's assassin. Blau and her friends attempted to warn Trotsky of the danger the young man posed, but he dismissed their appeals and was murdered in 1940.[108]

♦ After the end of the Spanish Civil War (1936–1939), which General Franco had won with the support of Nazi Germany and fascist Italy, tens of thousands of supporters of the Spanish republic fled to France. The Mexican government not only invited the Spanish refugees to come to Mexico but also accepted political refugees from Central Europe. Many of them belonged to various left-wing groups – social democrats, communists, Trotskyites. The different political affiliations often resulted in severe conflicts within refugee groups. After France fell, the Mexican consul in Marseille, Gilberto Bosquez, attempted to provide refuge for German and Austrian intellectuals by generously providing them with Mexican visas. Once in Mexico, the refugees developed a diverse cultural life. In 1941, the Heinrich Heine Club was founded as a cultural organization. ♦

Soon afterwards Austrian immigrants founded the "Acción Republicana Austriaca de México" (ARAM), in which all the political groups cooperated. In March 1942, it held a meeting called "Austria Will Rise Again"; moreover, the organization edited a monthly journal *Austria libre* and established regular cultural events. One such event is mentioned in Egon Erwin Kisch's book *Läuse auf dem Markt*:[109] "The Viennese laryngologist Dr. Leo Deutsch and Dr. Ernst Römer gave talks on music, the physicist Prof. Dr. Marietta Blau about the sun in Mexico." Some of ARAM's leading members, such as journalist Bruno Frei, were communists, a fact they did not publicize in Mexico. Among the members and supporters of this organization were Prof. Richard Volk, former director of the Vienna Lupus Sanatoriums; Carl Alvin, who had conducted at the Vienna State Opera between 1920 and 1938; the composer Marcel Rubin; the surrealistic painter Wolfgang Paalen, a friend of Diego Rivera; and Ernst Robitschek, a student of Max Reinhardt. Blau also became a member of ARAM. Because her contact with Albert Einstein was known among the German and Austrian refugees in Mexico, some people believed that she had been the famous scientist's secretary. Another story circulating was that she had come to Mexico on a boat together with the German writer Anna Seghers (1900–1983) and her son Pierre Radvanyi[110] (Fig. 17), who actually arrived in Mexico four years after Blau.[111] Blau was possibly mistaken for Trude Kurz, another Austrian physicist, who fled from France to Mexico during the war. Rumors of this sort found their way into some biographical articles on Marietta Blau.

The women in exile played a particularly important role. They were the ones who created a remnant of home-like atmosphere despite the poor living conditions in this distant land in which they now found themselves. But they also contributed to the intellectual life there and influenced scientific and cultural developments. In addition to Seghers and Kurz, this was true for the German actress Steffi Spira, the Czech journalist Lenka Reinerová, and the Viennese physicians Else Volk and Marie Frischauf-Pappenheim. Some of them were indispensable contributors to the activities of both the Heinrich Heine Club and ARAM. Gisl Kisch transcribed her husband's manuscripts and prepared his literary commentaries for publication. Thanks to Irma Römer, the Römer family home became a cultural center where refugees and Mexicans met. Trude Kurz organized a

Fig. 17: Anna Seghers with her children in 1943 (Archive Pierre Radvanyi).

women's committee among the Austrian refugees. She was a professor at the Autonomous University of Mexico (UNAM) and one of the first professors of computer science. Gertrude Duby-Blom gained fame as an ethnographer for her studies of the Lacandones, a native people living in Chiapas. Ruth Deutsch de Lechuga gained a reputation as a major collector of original Mexican popular art.

Anna Seghers asked Blau to teach her son Pierre Radvanyi physics. This boy later became a renowned professor of physics in France and held the position of research professor at the CNRS in Orsay, France (see p. 128).

On May 8, 1942, Mexico declared war on Germany and Italy. In August 1942, Blau gave her assessment of the war:

> For a while I was quite depressed, but today, after the landing of English and American troops in France and the more

favorable news from Russia, I am more confident again. Not that I would think even for a moment that the Germans could win, but I do fear that their final defeat might not come as fast as I had hoped and that it may cost even more victims. I would be happy to get personally involved in the war and stand behind a cannon or get shot as a human torpedo against a German ship, but they won't take me. Also, I would love to join the war industry, but here in Mexico they don't need me, and I can't go to the United States.[17, 112]

In 1943 Blau's hopes to continue her scientific research concentrated on Manuel Sandoval Vallarta, who started a cosmic-ray group at the UNAM. Indeed, some association with this group evolved, but it never led to actual employment.

Manuel Sandoval Vallarta was born in 1899 in Mexico City. Pursuing his great interest in mathematics and physics, he traveled to the United States to study at the Massachusetts Institute of Technology (MIT), where he specialized in theoretical physics and received his Ph.D. at the age of twenty-five. After post-graduate studies in Berlin and Leipzig under the guidance of Einstein, Planck, Schrödinger, Heisenberg, and Debye in 1927/28, Sandoval Vallarta was appointed professor at MIT. He often visited Mexico while at MIT and encouraged the development of a national scientific research program. His own research fields included the theory of electric circuits, quantum mechanics, general relativity, and, from 1932, cosmic rays. He studied the nature and composition of cosmic rays and the effect of the geomagnetic field on their constituents and was considered one of the authorities in the field.

In 1943 Sandoval Vallarta returned to Mexico City and worked part time at the UNAM while continuing to teach at MIT. He was chairman and a member of the Comisión Impulsora y Coordinadora de la Investigación Científica from 1943 to 1951. Evidently, he would have been a competent partner for Blau in cosmic-ray research. Why closer cooperation never developed between the two is unclear. Sandoval Vallarta is said[92] to have put forth some effort to help Blau continue her studies on cosmic-ray-induced nuclear reactions, but nothing tangible resulted. In the Polytechnical School, the administration and the bureaucracy began to dominate academic life and scientific investigations. These people had a rather limited understanding of Blau's work and placed little value upon it, and as a person, she was always considered second class.[92]

The ARAM arranged for Radio Gobernación to broadcast *Voice of Austria* once a week. In one of these programs, on January 29, 1943, Blau gave a talk on Stefan Meyer.[113] The talk took place only a short while after Blau had suffered a severe loss: After a painful illness, her mother died of liver cancer[11, 114] on January 20, 1943 (Fig. 18). Two weeks later, Blau moved to another apartment. When she contracted typhus in the same year, her own health was also affected.[106]

Fig. 18: Florentine Blau's gravestone (Photo courtesy of H. Yee Madeira).

Upon her departure from Mexico a year later, the Instituto Mexicano Europeo de Relaciones Culturales arranged a party in her honor at which the French author Jules Romains[115] gave the farewell speech.[116]

THE VIENNA RADIUM INSTITUTE: 1938–1945

The policies of Nazi Germany were rigorously carried out in occupied Austria, including at the Radium Institute. The director Stefan Meyer and his vice director Karl Przibram had been removed from office but were still initially allowed to work at the institute, but they were later banned after one of the scientific personnel threatened to denounce them. According to Przibram, "Meyer withdrew to his cottage in Bad Ischl, where he survived the National Socialist regime thanks to his reputation and his daughter's careful diplomacy."[117] Przibram also survived underground in Brussels. In 1943, the Radium Institute, together with the Second Physics Institute, formed an Institute for Neutron Research (Vierjahresplan-Institut für Neutronenforschung), financed by the Reich Office for Economic Development (Reichsamt für Wirtschaftsausbau) in Berlin. Georg Stetter[A9] became head of the neutron-research institute. He had been a staff member at the Second Physics Institute and was also a corresponding member of the Academy of Sciences. Gustav Ortner,[A10] a former university assistant at the Radium Institute since 1924, was second in charge. Stetter and Ortner had worked together before 1938 and combined an ionization chamber with a tube voltmeter, thereby creating an objective method for detecting particles emitted in nuclear reactions. Stetter had applied for membership in the NSDAP in 1933; his application had not been processed because of the illegal status of the NSDAP at that time. Nevertheless, from 1937 on, he passed as a member of the NSDAP.[118] The institute's scientific research concentrated on nuclear fission, which had been discovered in Germany in 1938; the fission of various actinide nuclides was then proven at the Vienna Radium Institute, and the first precision measurement of the energy produced in uranium fission was performed using its neutron source.

By removing Stefan Meyer and also "non-Aryan" professors Ehrenhaft, Przibram, and Kottler from the physics institutes, two positions for associate professors and two positions for full professors became vacant. Besides the non-Aryan professors, Hans Thirring was also removed because of his ardent opposition to the National Socialists and his unwillingness to compromise with them.

In Germany it was customary to replace Jewish professors with lecturers from other universities. Therefore, physicists from Germany became interested in the vacant positions in Vienna. However, there in Vienna the intention was to fill the posts with Austrian professors.

For years Kirsch,[A5] Ortner,[A10, 119] and Stetter[A9, 118] had been members of the NS teachers' association, which was considered part of the NSDAP from 1937 on, and they therefore claimed long-term membership in the party in 1938. When the experimental physicist Hans Bartels from Hannover applied for one of the vacant posts, it was stated that Kirsch, Stetter, and Mattauch[120] were morally entitled to these positions. In fact, Stetter became head of the Second Physics Institute, Kirsch of the Third and later the First Institute, and Ortner took over Meyer's position as director of the Radium Institute of the Academy of Sciences in 1939. Mattauch succeeded Lise Meitner at the Kaiser-Wilhelm-Institut in Berlin-Dahlem.

Hertha Wambacher, Blau's former collaborator, who was said to have been a member of the NSDAP since 1934,[121] developed a close relationship with Georg Stetter, presumably out of the desire for personal appreciation, and to seek scientific guidance. Together with him, she continued to investigate nuclear reactions induced by cosmic rays by means of photographic plates and published the results between 1939 and 1945, partly by herself[122, 123] and partly with Stetter[124, 125] and one of his thesis students.[126] By September 1938, she had already presented a paper titled "Multiple Disintegration of Atomic Nuclei by Cosmic Rays"[127] at the meeting of physicists and mathematicians in Baden-Baden. In the academic year 1939/40, she applied for *Habilitation* on the basis of the work "Nuclear Disintegration by Cosmic Rays in Photographic Emulsions"[122] and received the degree of Ph.D. habil. After having proven her abilities as a teacher to Stetter's satisfaction in September 1940, she was entitled to teach physics by the Reich Ministry of Education, Science and Adult Education,[128] as proposed by the dean of the faculty and she taught classes in the following years.

MOVE TO THE UNITED STATES – WORK IN INDUSTRY

In 1944, about a year after her mother had died, Marietta Blau obtained permission to immigrate into the United States. The timing suggests that she had persevered in Mexico in order to avoid having to uproot her mother again, and only after

Florentine Blau had passed away did she apply for a permit to immigrate into the U.S. Indeed, getting an immigration permit for the U.S. would have been possible earlier since the quota was not filled after 1938.

> ♦ Consideration for their mother seems to have been quite decisive in Mrs. Blau's children's lives. Marietta's brother Otto only married at well over fifty, after her death.[129] ♦

In May 1944, Blau moved to New York City, where her younger brother Ludwig was living, and took a job in industry, which she held for two years, with the research department of the International Rare Metals Refinery Inc. She was later employed at the Canadian Radium and Uranium Company. At that time, she developed various devices and filed patents for them individually and together with colleagues and published papers on metrology in radioactivity in scientific journals.[47] In a letter to Stefan Meyer, Blau wrote:

> I could also write a chapter on the industrial application of radioactivity (old [in the sense of naturally radioactive preparations, as opposed to man-made radioisotopes produced in nuclear reactors]). Unfortunately, I have a lot of experience in this field and own a whole series of patents and so on [presumably instrumentation for which no patents were registered], which in most cases make use of radioactivity and electronics, where I have also acquired experience.[130]

In "the large oppressive town,"[106] one could at least attend concerts. Marietta's situation deteriorated in summer 1947, when the company for which she worked merged with the Gibbs Manufacturing and Research Corporation. She was transferred to Janesville, a town of twenty-five thousand in Wisconsin. Her discontent is reflected in a letter to Meyer:

> Please don't be angry with me for not writing, but I have had difficulties and unpleasant experiences for quite a while. I do not want to complain but cannot bring myself to write cheerful letters either.

> Circumstances are not too good here at the moment, and the firm for which I work has merged with another, and I am in a little town in the Middle West. Anyone familiar with Sinclair Lewis will know what that means. But I would even accept this if only the work were just a little bit interesting. Since all cultural diversion is lacking, there is only nature left to enjoy, which I do when the temperature (96° F) and 99% humidity allow ... I will try to get away from here, but letters to this address ... will still reach me.[131]

Blau's aversion to "the scientific and particularly the social environment"[130] in Janesville was as troublesome to her as the fact that at that time she did not drive and hence could hardly visit towns further away. On top of all that, the hot and humid climate took a toll on her fragile health.

It may be worth considering to what extent the hardships Blau had to face here were aggravated by the fact that she had to master them by herself. Whatever her reasons were for not marrying, her status as a single woman – apart from the emotional privation – certainly complicated her life in social settings. Being single is something to which little value is attached even today, and it was even more so in the U.S. society of the 1940s to 1960s.[11] The assumption that men are the ones who count and that women exist only in relationship to them was accepted sublimally as a matter of course, something not even questioned by women until recently.

Before long, Blau started writing to colleagues, universities, and firms in the hope of returning to the scientific research she had sorely missed for nine years. Employing her in a government program was difficult because she was not yet a U.S. citizen. The president of the Canadian Radium and Uranium Company also pointed to this fact in a letter to Blau, in which he expressed both his understanding of her plight and his fervent desire to retain a person of her high ability in his company.[132]

Finally, she was offered a job at Columbia University in New York as a scientific staff member. She accepted the two-year position beginning January 1, 1948, at an annual salary of $7,200. Cheerfully she told Meyer: "It's not quite sure yet but already very likely, that I may take part in a government project starting next year. As I am not yet a U.S. citizen, I will probably not work directly there but will have a laboratory at Columbia University. I am so happy about it, but now I have to fight battles with my present employers so that they will let me go."[101]

♦ While seeking new employment, it was convenient for Blau to use her brother Ludwig's home as a base. She moved to an apartment of her own on West 173rd street in New York City at the beginning of 1948. This is in Washington Heights, the district where many German and Austrian refugees were living at that time. It was the New York equivalent of Leopoldstadt, where Blau grew up. About a year later, she moved to Riverside Drive, close to both the Hudson River and Columbia University. ♦

BLAU'S FIRST RESEARCH PROJECTS IN THE UNITED STATES: COLUMBIA UNIVERSITY

♦ Columbia University in New York City was chartered as King's College in 1754 by George II of England; it is the oldest institution of higher education in the state of New York and the fifth oldest in the United States. After the American Revolution, the college re-opened in 1784 under the new name "Columbia." Towards the end of the nineteenth century, Barnard College for women, the medical school, and the Teachers College affiliated with Columbia, and a graduate school was formally established. Shortly thereafter, the name Columbia University was authorized, and the university moved to the spacious Morningside Heights campus, designed as an urban academic village. The Columbia Physics Department played a significant role in several of the developments of modern physics through a number of Nobel laureates who worked there either as visiting scientists such as Hendrik A. Lorentz,[133] and Max Planck,[134] or as faculty members such as Enrico Fermi,[135] Isidor Isaac Rabi,[136] Polykarp Kusch,[137] and Willis Lamb.[138] ♦

At Columbia, Blau was again the only female scientist.[23] Her task was to investigate and possibly improve the application of photographic emulsions for the study of high-energy particles. These studies were extremely important for the work the Atomic Energy Commission was conducting at the Brookhaven National Laboratory.

> ♦ In the 1940s, nuclear research started to focus on work with machines that accelerate charged particles, such as electrons, protons, and ions, to high energies. The first accelerator called a cyclotron, was built in the 1930s, and had a diameter of only nine centimeters. After the war, particle accelerators became larger and larger, thus increasing the particle energy that could be achieved. Famous examples of such machines were the Cosmotron and the Bevatron. The accelerating machines could produce energized particles with much higher intensity than cosmic rays. When a particle with such high energy collides with a nucleus, it causes the disintegration of the nucleus, whereby new particles can be formed. In this way, various particles, such as different types of mesons, were discovered. ♦

The application of nuclear emulsions for detecting particles was provided for in the projects at the Nevis Cyclotron of Columbia University, which was then being built north of New York City, and in the investigation of materials formed in fission reactors. Since Marietta Blau (Fig. 19) was a well-known pioneer in this field, she was entrusted with designing the corresponding research program. The photographic method, which she had originally taken up and developed for detecting particles in low-energy nuclear reactions, had already led her to cosmic-ray studies before her emigration. She had explored the reactions induced by cosmic rays and the elementary particles produced in them. Now the application of nuclear emulsions drew her to the large accelerators that were just being constructed and that competed with cosmic rays as the only source of gathering knowledge about elementary particles.

Fig. 19: Marietta Blau in the late 1940s (Berlin-Brandenburgische Akademie der Wissenschaften, Archive).

How much she enjoyed the new assignments is noted from her correspondence with Stefan Meyer:

> I'm working at Columbia University and am extremely happy about it. Since I have been concentrating on technical things for years, I have to get acquainted with the new physics, but I am good friends with the meson already and know everything that has been written about it. I even got to see it on plates, but still it is somewhat mysterious.[139]

At first Blau had space in the Pupin Physics Laboratory of Columbia University in New York City at her disposal. Pupin Hall was later designated a national historic landmark in recognition of the atomic research undertaken there by Columbia's scientists beginning in 1925. Martin Block, later a professor in Illinois, was a participant in the construction of the cyclotron (see

p. 130). Seymour J. Lindenbaum, who planned the research at the cyclotron, reports that Blau built up an emulsion lab in a short time and trained the staff herself (see p. 130). In summer 1948, she wrote to Hans Thirring in Vienna: "I'm very glad to have moved from work in industry and teaching in Mexico to research again. But the stress of the long, difficult years weighs upon me so that I cannot enjoy the new work as much as I should."[140] In the same letter, she mentioned collaboration with an "old acquaintance" from Vienna, Robert Pollak-Rudin. Together with him and Lindenbaum she developed a semi-automatic method[B64] for evaluating emulsion results and was thereby able to process the huge amount of data that would be accumulated in the cyclotron experiments. In her methodical studies together with her students, she treated processing techniques of nuclear emulsions as well as parameters that could be used for characterizing and identifying tracks.

Now that she was living in the United States, Marietta Blau's private life flourished as well since she had contact with old friends from Vienna, among them Herta Leng, Elisabeth Rona, and Peter Havas.[141]

After her two-year contract expired, Blau was hired by the Atomic Energy Commission at the Brookhaven National Laboratory (Long Island, New York).

♦ The beginnings of this institution date to 1946 when representatives from nine major eastern universities – Columbia, Cornell, Harvard, Johns Hopkins, MIT, Princeton, Pennsylvania, Rochester, and Yale – formed a non-profit corporation to establish a nuclear science facility, and thus Brookhaven National Laboratory was created. It was intended to promote basic research in the physical, chemical, biological, and engineering aspects of the atomic sciences and also to design, construct, and operate large scientific machines that individual institutions could not afford to develop on their own. The laboratory was to resemble a university to the greatest possible extent. In the course of its history, it was the home of several research reactors, numerous particle accelerators, and other research machines. ♦

When Marietta Blau started at Brookhaven in 1950, "some of the high-energy machines were already completed."[47] The emulsion method was used to investigate "the reactions that take place when high-energy protons accelerated with these machines collide with the light and heavy elements which constitute the emulsions as well as with free hydrogen particles which are abundant in the emulsions."[47] Her employment became possible only after she had obtained U.S. citizenship. In the midst of the McCarthy era, individuals suspected of communist collaboration or sympathy, primarily scientists and artists, were closely scrutinized and investigated. Despite her former membership in the Acción Republicana Austriaca de México, in which communist officials had been active, Blau was at last considered politically unobjectionable by the authorities although her contact with the well-known communist Anna Seghers in particular seems to have caused some problems.

In 1950, when Blau finally had access to the most modern and expensive research techniques, two events that reflected a lack of appreciation for her work before 1938 must have been especially disappointing for her. The Nobel Prize in Physics was awarded to Cecil F. Powell[A11, 142] for the development of the photographic method for particle detection, and her former collaborator Hertha Wambacher was widely praised for her work at the time of her death.

As far as the Nobel Prize goes, it must have been gratifying on the one hand for Blau to see the achievements in the field to which she had contributed greatly held in high esteem but on the other hand disappointing that her pioneer work went unrewarded. In fact, Erwin Schrödinger (Fig. 20) had also proposed Blau and Wambacher for the Nobel Prize in Physics in 1950. In his letter to the Nobel Committee,[143] which is based on the two women's publications before 1938,[B38, B39] he pointed out how important their method was in exploring cosmic rays in general and that, in particular, they had been the first to interpret the stars correctly as atomic disintegration induced by cosmic rays. Powell, for his part, had started to work on the photographic method at Bristol University in 1938 and used it for registering tracks of cosmic rays. Furthermore, he observed the tracks of recoil protons to determine the energy of neutrons produced by a Cockcroft gene-

Fig. 20: Erwin Schrödinger around 1950 (Archive, Österreichische Zentralbibliothek für Physik, Vienna).

rator he had constructed himself. To improve emulsions, he had experimented with high contents of silver bromide, on which he first collaborated with Ilford and later also with Eastman-Kodak, whereas Blau had worked to increase layer thickness and had also cooperated with Ilford. In 1947, Powell discovered tracks of π-mesons in photographic plates exposed to cosmic rays, which was partly the reason he was awarded the Nobel Prize. He co-authored the book *Nuclear Physics in Photographs*.[144]

Blau's achievement in developing the photographic method was never mentioned, either by the Nobel Prize committee or in Powell's speech. In 1970, he wrote in an autobiographical note that he only started to utilize this method after he had learned from Walter Heitler, who knew Blau personally, about Blau and Wambacher's papers and found the photographic method to be much superior to the Wilson cloud chamber for his experiments. But this statement was made long after he had received the Nobel Prize.

♦ Many people, who are familiar with Blau's scientific achievements, feel that "she was strongly entitled to share Powell's glory"[145] and that the Nobel Prize should have been shared by the two of them. (The award was given after Wambacher's death.) The 1950 protocols of the Nobel Committee reveal to what extent Powell's nominations (fourteen) exceed the single one of Blau and Wambacher both in number and thoughtfulness. Here is an excerpt from Schrödinger's nomination:

I herewith propose that this year's Nobel Prize for Physics be given to Marietta Blau and Hertha Wambacher for experimental work conducted and published collaboratively. *I believe that they were the first to attempt to expose sets of photographic plates for several weeks at high altitude and then develop them in order to find traces of cosmic radiation particles; I also believe* that they were the first to discover on those plates those famous "stars" and to interpret them correctly as explosions of atomic nuclei produced by high-energy cosmic-ray particles. Their first preliminary publications *seem to have been Nature* 140, p. 585, 1937; *Wiener Anzeiger* p. 215, 1937 ... I therefore consider Marietta Blau and Hertha Wambacher to be worthy candidates for the Prize, *if it can be established beyond a doubt* that they were the first to use this method or to discover the "stars." *I believe* both of these to be the case. *I am told*, that Marietta Blau is presently at Columbia University, Pupin Institute, New York, N.Y. About Hertha Wambacher *I know nothing*. For all I know *either of them may have married* and thus changed their names.[143] (Italics by authors.)

The comparison of the support of Powell and Blau confirms Walter Thirring's[146] later assessment that the obstacles to Blau's career were not only the fact that she was discriminated against as a minority but also that she lacked an effective lobby.[147] ♦

As is seen from the protocol[148] released in 2001 by the Royal Swedish Academy of Sciences, the referee, Axel E. Lindh,[149] was of the opinion that the achievements of the two women were not sufficient to be granted the Nobel Prize. He rejected their nomination by arguing that the idea of applying the photographic method to the exploration of cosmic rays had originally come from Eduard G. Steinke, which Wambacher had mentioned in her talk at the meeting of physicists and mathematicians in

Baden-Baden in September 1938,[127] and that it was thanks to the scientific staff of Ilford Ltd. that Blau and Wambacher had been able to make the stars on the photographic plates visible.

Because of the fact that Blau had had to spend over twelve years far from home, most of them removed from significant research opportunities, it seems ironical that she now had to watch how those whose fates had been more favorable gained international recognition for the photographic method for which she had laid the foundation. She probably could have emigrated to England in 1939, like her brother Otto. After the pogrom in November 1938 ("Night of Broken Glass"), immigration regulations were somewhat relaxed and the family owned the Josef Weinberger publishing company, which had a branch in Great Britain. However, the idea that she could have joined the Bristol research group and participated in the discovery of the π-meson is wishful thinking.

When Marietta Blau visited Great Britain after the war, she considered living in England attractive, as is seen from a letter she wrote to her niece during a visit in London: "Life in England appears to me much more beautiful than in the States ... The landscape, even in the fog, is just so lovely and peaceful, and one gains constant delight from the beauty of the buildings and the culture they represent. People seem to be much closer to life, more original and interesting than the average American. I do think that everyday life – especially for women – is much easier in America than it is here, but in the long run and in professional life everything is more human and simpler here."[150]

In her treatise[B79(1)] on photographic emulsions ten years later, she described the developmental work in those companies from a purely scientific point of view:

> The studies were much improved by the availability of new emulsion types containing higher concentrations of silver halides with which denser and hence better defined tracks could be obtained. These emulsions were first produced by Ilford, later also by Kodak and Eastman-Kodak. However, until 1948 the photographic method was restricted at best to the detection of particles at forty per cent of the speed of light; emulsions available then were not sufficiently sensitive to

register particles of higher velocity and hence lower specific ionization [the number of ion pairs per unit length]. In 1948, Kodak Ltd., England, and shortly thereafter Eastman-Kodak and Ilford, succeeded in producing so-called electron-sensitive or minimum-ionization emulsions with which all charged particles could be registered, irrespective of velocity.

POST-WAR AUSTRIA AND THE PHYSICS INSTITUTES IN VIENNA AFTER 1945

After Vienna's liberation by the Red Army, representatives of all political groups except the Nazi party, i.e., social democrats, conservatives, and communists, formed a provisional Austrian government on April 27, 1945. The Second Austrian Republic was established while fighting continued in some parts of the country. The Soviet Union was the first to recognize the new government, and a few months later the United States, Great Britain, and France followed suit. Thus, in contrast to Germany, Austria regained a certain degree of sovereignty soon after the war, even though the country remained occupied by allied troops. Austria was divided into four occupation zones, as was the capital, Vienna. The images of the four allies patroling together in one vehicle (*"Die Vier im Jeep"*) are legendary. But the government and the democratically elected parliament had the authority to decide most internal issues with little interference from the occupying powers. And thus, the laws aimed at removing former members of the NSDAP from positions in civil service and the universities were the result of decisions made by the Austrian government.

Hertha Wambacher had been a staff member at the Second Physics Institute during World War II, and when Austria was established anew, she was removed from the university for having contributed to Austria's demise, as were all other individuals who had been members of the Nazi party before 1938. Thus Kirsch, Stetter, and Ortner were fired along with Wambacher. By the end of April 1945, she was forced to go to Russia as was Ortner; she is said to have returned only in 1946,[151] whereas Ortner came back to Austria in late summer of 1945. Shortly after her return, Wambacher contracted cancer,[151] although she was able to continue to work in a research laboratory in Vienna.[152] Blau had started corresponding again with Meyer and Karlik after 1945, but there is no indication she did so with Wambacher, who had once been her closest collaborator. When Blau bade her farewell in March 1938 for what was to be four months, it turned out to be for the rest of their lives: Wambacher died of cancer on April 25, 1950, after considerable suffering. Obituaries were published by Georg Stetter alone[153] and together with Hans Thirring,[154] and also by Georg Wagner[155, 156] and Gustav Ortner.[157] The obituaries written by Stetter and by Wagner create the impression that Wambacher was the leading originator in developing the photographic method and discovering the disintegration stars. The fact that she had first examined the plates exposed at Hafelekar under the microscope for their team was used to her advantage. However, the details that Wambacher had been Blau's student, that the discovery of the stars had been possible only because of their common preparatory work, and that Blau had always played the leading role on this team, are inadequately expressed, if not to say suppressed.

After the war, Stefan Meyer was formally reinstated as director of the Vienna Radium Institute although he never actually served in that capacity again. Directorship of the Radium Institute was provisionally transferred to Berta Karlik after the war and permanently in 1947. She had to face many difficulties in this position:[158] The Institute was filled with pieces of broken glas as well as dust and debris blown in during the bombing of the surrounding areas. The equipment and radioactive preparations had been brought to the countryside in order to protect them. Gas and electricity were turned on only for a very short time per day, and fuel was scarce. Apart from these hardships, the Institute's reemergence in research was hampered by the fact that Austrian scientists had been barred from access to international scientific literature for a long time because of the policies of the Nazis and the disruption of war. Karlik now turned to the Institute's emigré friends for help. Among others, Marietta Blau sent physics books and journals to the Radium Institute.

Blau had written a report[B61] on the discovery of the stars in photographic emulsions for the publication celebrating the fortieth anniversary of the Radium Institute shortly before Wambacher's death. In it, she describes their collaboration in a fair and complete manner. The article concludes with the typical acknowledgment of the continuous encouragement "from our dear Stefan Meyer and the beneficially harmonious atmosphere at the Vienna Radium Institute." Stefan Meyer himself did not live to take part in his Institute's jubilee; he had passed away by the end of 1949. As a result, the third edition of the work *Radioaktivität* by Stefan Meyer and Egon Schweidler, which Meyer had been planning since 1947 and for which he had arranged[159] contributions by Marietta Blau, Berta Karlik, Karl Przibram, and even Lise Meitner, was never published.

Ortner and Stetter had protested their dismissal and maintained they had merely given the Nazi teachers' association donations for charitable purposes and had not applied for membership in the NSDAP before March 1938. However, in 1938, the representative of the Nazi teachers' association had confirmed that they were members of this banned organization. Stetter defended his actions in a letter stating that he had never derived any personal advantage from his membership in the party. Regarding the events in 1938, he wrote: "In any case, one knew only about the lofty ideals, and it was only later that one learned of the actual activities of these incompetent and shady creatures. My opinion became increasingly critical and finally negative."[118] Even in this attempt to exonerate himself, he never commented on the National Socialists' crimes. He tried to prove that his behavior during the National Socialist era was appropriate and that he had even tried to protect Stefan Meyer, Hans Thirring, Eduard Haschek,[160] and others as far as possible. Indeed, the dean's office had already obtained a letter[118] from Hans Thirring in January 1946 stating that Stetter's behavior towards him had always been proper and helpful, even though Stetter had been aware of Thirring's political position. In his letter, Thirring also stated that Stetter had never denounced any of the obvious opponents to the Nazi regime at the Institute during the Nazi era, reflecting the remnants of his loyalty to his colleagues. Stetter himself explained further: "Regarding the Jewish question as it relates to science, I expressed my position – at this time, students expected their teacher to make a statement – that the Jewish way of conducting science and research differed from the German way but that it was ridiculous, however, to doubt or even to suppress new knowledge about the field of physics merely because it originated from a Jew."[118] In fact, Blau's contributions to the field had been recognized inasmuch as her collaboration with Wambacher formed the basis of the further work of Wambacher and Stetter. Blau's part in the authorship is, however, given only fleeting credit in Stetter's papers with Hertha Wambacher, as well as in his obituary for her.

In October 1950, the area-government office in Zell am See decreed that Stetter's name be removed from the Nazi party registration list, and thus he was no longer in violation of the law of 1947 forbidding the NSDAP. The purported reason for removing his name was that he had never been issued a Nazi party membership card so that according to the party statute he had never been a member and the confirmation of his membership was to be viewed as a favor to him at the time. That same year, Stetter's credentials to teach were reinstated.

Ortner never wrote such an extensive justification,[119] but in a letter to the dean, a female physicist reported that Ortner had helped her finish her studies and obtain a job although according to the Nazi (Nuremberg) racial laws she was of mixed blood (a *Mischling*). Like many other former Nazis, Ortner taught abroad (at Cairo University from 1950 to 1955); his authorization to teach in Vienna was reinstated in March 1955.

When a successor for Ehrenhaft, who had returned from the United States after the war to teach again at the University of Vienna, was to be selected in 1952, the committee responsible for the appointment ranked the German Wolfgang Paul first and Stetter second. The election was by a show of hands with two votes cast against Paul and nine against Stetter. When Paul turned down the position and went to the University of Bonn instead, Stetter was appointed. The Federal Chancellory agreed to the appointment, even though there were concerns about Stetter's having his name struck from the registration list based on inaccurate statements regarding his NSDAP membership. But the weight attributed to these doubts was not sufficient to consider his appointment objectionable from a legal standpoint, especially since the Ministry of Education placed a high value on it.[118]

When the Western Allies began to regard the USSR as their primary opponent as the Cold War intensified at the end of the 1940s, the attitude of the Austrian government also shifted. The pursuit of Nazis became a lower priority; and since former NSDAP members were even allowed to participate in the election of 1949, the political parties began to court them as well. This fact may explain why the rehabilitation of even ardent Nazis proceeded so easily after 1950.

On May 15, 1955, the Austrian Treaty of State was signed, granting Austria full sovereignty.

BLAU'S SCIENTIFIC RESEARCH IN BROOKHAVEN AND MIAMI

When Marietta Blau started working at the Brookhaven National Laboratory (BNL), Leland John Haworth held the position of director. Born in 1904, he completed his education in Indiana and Wisconsin and had been assistant and later full professor of physics at the University of Illinois between 1940 and 1947, although he was on leave to do work for the war effort for most of this time. He became director of the BNL in 1948 and remained in this position until 1961.

In Blau's work at the Brookhaven Cosmotron (Fig. 21), π-mesons, in addition to protons, were increasingly used as incident particles. Her research yielded important results, such as the confirmation of the detection of additional-meson production in meson reactions[B68–B71] and the production of hyperfragments (unstable nuclei with a heavy particle [hyperon] in place of one of the neutrons).[B73]

Blau was employed as a senior scientist, and there were two men at BNL who were in the same rank. One was Maurice Goldhaber, born in 1911 in Lemberg (then Austria, now Lviv, Ukraine), who had also pursued his academic career at the University of Illinois between 1938 and 1950, and then moved to the BNL. He and Blau became good friends. He succeeded Haworth as director in 1961. Edward Oliver Salant, on the other hand, was an American who had joined the BNL as a senior physicist right at its opening in 1947 while a professor at New York University. Even though they were not particularly fond of

Fig. 21: The Cosmotron at Brookhaven National Laboratory (Photo courtesy of Brookhaven National Laboratory).

each other, Blau and Salant certainly shared many common research interests, in particular the reactions of fundamental particles from cosmic rays and from the new accelerators. They published just one paper together.[B67]

In fact, the atmosphere at BNL appears to have been quite competitive. Blau told Karlik: "I'm working on cosmic-ray problems and, therefore, started several investigations at the Columbia Cyclotron, but I doubt if I will be able to complete them, since others, in particular Bernardini,[161] who is a guest professor at Columbia and incredibly ambitious, pounced upon them."[162]

Brookhaven National Laboratory is situated on the Eastern part of Long Island, New York (Fig. 22) at some distance from the Long Island Railroad. Even in 1980, people quipped: "Public transportation to the BNL is not bad, it simply does not exist."[163] Thus Marietta found herself once more in a rural area, isolated from large cities where she could have attended cultural events.

Fig. 22: Map of Long Island on a greeting card Blau sent Karlik for Christmas, 1954 (Archive Radiumforschung, Archive Österreichische Akademie der Wissenschaften).

She found a place to stay in Patchogue, a village about six miles southeast of the lab. At long last, she received her driver's license and bought a car, but she practiced a rather peculiar manner of driving: She stopped when and where she pleased, infrequently at red lights and stop signs.[164] Once she found herself on the railroad tracks between closed gates. Luckily, the train stopped.[165]

When Blau traveled to Europe in 1953, she was invited by Berta Karlik to visit Vienna. At this time, she would have required a special permit to come to Vienna since the eastern part of Austria was still occupied by Soviet troops (cf. p. 80). In England, Blau (Fig. 23) also went to see Powell in Bristol as she relayed in a letter to Karlik:

Thank you very much for your friendly letter and for your invitation. Naturally, I was thinking of coming to Vienna, but the difficulties of getting a permit from Washington to do so were too great and would have taken too much time. Besides,

I was just so tired before I left that I decided to spend the vacation quietly with my brother [Otto]. Now I feel that I have somewhat more energy and would like to come, but my vacation is over, and it would not by very wise to extend it, as the Cosmotron is operating again. It is true that I have prepared everything for that, but since the first experiments are already running, I have to go back now. Chose the worst time for England since I am used to overheated American buildings and I came to completely inadequately heated houses and am terribly cold, probably because of the dampness. But otherwise, it is very nice – particularly to be with my brother and old friends, to partake of family life, and not to have to worry about anything. Naturally, I enjoy the beautiful old buildings and museums, the tasteful apartments, the people who are not under constant stress, and even the countryside covered by fog. I am just now in Bristol in Powell's institute and see much of interest. I am staying for a few days and then I go back to London and one day in Cambridge, and then on the twenty-third, I fly back to New York. Last year my health was not too good, and probably I had worked too much and on top of that on projects that were not even interesting. But I hope that at least a paper on cosmic rays and a paper on neutron and π-meson stars in light elements will be published soon.

Is it true that Hess is celebrating his seventieth birthday?[166]

In January 1955, Hans Thirring nominated Blau for the Nobel Prize in Physics for "her pioneering work in the study of high-speed atomic particles made visible in photographic plates" which formed the basis of the method "that is presently one of the most important for the study of high-energy processes of cosmic rays and meson production."[167]

The proposal was not accepted. Erwin Schrödinger nominated Blau for the Nobel Prize in Physics once more before 1960[168] but without success. (The protocols from that period have not yet been disclosed.)

88 Marietta Blau – The Woman

Fig. 23: Marietta Blau in the 1950s (Archive R. and L. Sexl).

In 1955, Blau became increasingly dissatisfied with her employment at Brookhaven. Not all the projects that were difficult were of interest to her.[166] She felt tired[169] – remarks about "having been ill," "not feeling all that well," and "probably being overworked" had appeared in her letters repeatedly for years – and the atmosphere in the laboratory was not optimal.[170] Now at the age of sixty-one, she wished she "had the time to sit down and write, in part about physics and in part my memoirs and experiences."[162] She did not consider retiring from her work in science: the results of new research still fascinated her,[169] and retirement would have left her with too little money to live on. Of the thirty-five years she had worked, only those at Columbia University and at Brookhaven Laboratory entitled her to a pension. She took a leave of absence from the lab and accepted a position as an associate professor at the University of Miami, Florida, initially for just one semester.

♦ The University of Miami, a private university located in Coral Gables, was chartered in 1925 by a group of citizens when the community grew rapidly during the South Florida land boom and an institution of higher learning was deemed necessary. However, the institution was adversely affected by the collapse of the real estate market and the national economic depression in the following years. Around 1930, the university comprised schools of liberal arts, music, law, business administration, and education. In the 1940s, a graduate school and the schools of marine science and engineering were added. In the mid 1950s, total enrollment was about eleven thousand. New facilities and resources were aimed at increasing the research productivity of the institution, and doctoral programs were added in many fields. ♦

In February 1956, Blau drove by herself from New York to Miami. She was "deeply touched and saddened," to see how blacks were openly discriminated against in North and South Carolina.[170] She was thankful that such prejudice was not so

Fig. 24: Institute building in Miami, where Marietta Blau's laboratory was located, as seen from Santander Avenue (Archive Arnold Perlmutter).

pronounced in Florida, where the mild climate and the rich fauna and flora reminded her of Mexico. Nevertheless, she longed for her native country: "Despite all the beauty, I am frequently homesick and think of the wonderful days back home in Vienna."[170] In Miami, she initially taught various topics related to physics. When she decided to stay beyond one semester, she bought devices and laboratory equipment using funds from U.S. Air Force grants.[47] (The institute building in which Blau's laboratory was located can be seen in Fig. 24.) Because she was so well versed and confident in her abilities in science, she was soon collaborating productively with several colleagues and directed numerous student research projects. She motivated young Arnold Perlmutter,[A12] who had come to Miami shortly before her as a university assistant and who was working in solid-state physics, to join her research program in particle physics. The improved experimental equipment aided the determination of the ionization parameter, the characteristic quantity for particle identification. The particles now being investigated were antiprotons and negative π- and K-mesons. Collaboration with Perlmutter yielded a series of joint publications.[B75–B78, B80] Moreover, Blau became a close friend of Perlmutter (Fig. 25) and his family; it was as if she were one of them (cf. p. 132 and p. 181).

In Miami, Blau also met Fritz Koczy, a former colleague from the Vienna Radium Institute, who also worked at the University of Miami. Among her students was Sylvan Bloch, now professor

Fig. 25: Arnold Perlmutter around 1960 (Archive Arnold Perlmutter).

emeritus from the University of South Florida in Tampa, who remembers Marietta Blau vividly (see p. 136).

After the bloody suppression of the Hungarian revolution at the end of 1956, Blau invited a Hungarian girl, "a relative of friends of mine," to live with her and finish her studies in Miami.[171] In Mexico as well Blau had had a woman live with her and her mother. What Marietta offered generously may have been accompanied by the desire to see her kindness appreciated and reciprocated. In several cases, however, the persons who accepted her generosity simply took advantage of her and did not in the least think of giving back for what they had received. It is most likely due to these experiences that years after her death her brother Otto characterized her life as "rich in disappointment."[172] It seems that "looking out for the next pogrom"[29] was not so much the expression of her Jewish identity as it was of her mental disposition.

In 1957, Blau rented a little house[173] which shortened the distance to work. At that time, health problems began to plague her, primarily a heart condition and cataracts.[1]

At Erwin Schrödinger's request, Blau was awarded the Leibniz Medal of the German Academy of Sciences in Berlin in 1957 for having first observed disintegration stars and for developing the method of detection. Blau was at first jubilant but then had to inform Schrödinger that, as a U.S. citizen, she was not allowed to accept this honor because it originated from the Academy in East Germany, with which the United States did not have diplomatic relations. In order not to create problems for Blau, the request to grant her the medal was withdrawn for the reason that the guidelines for the award stipulate that it not be given to an active researcher. Even though none of these circumstances changed, she was granted the medal two years later, again at Schrödinger's request, but had to reject it this time as well by order of the State Department.[174]

In 1959, Blau fell on the garden stairs of her house and broke her left arm, which required surgery and an extended leave of absence. During her convalescence, she remained in contact with her lab while being looked after by her colleagues, their wives,

Fig. 26: Chien-Shiung Wu (Wu Chien Shiung Education Foundation, Archive).

her students, and neighbors "in a caring manner."[175] She had to pay for the medical bills herself. Her life became trying because of the injury to her arm and because her poor eyesight prevented her from driving.[175] At that time, she had already decided to return to Vienna. She needed an operation on her eyes, which she was advised not to have done in Florida so shortly after the surgery on her arm and which she could more easily afford in Europe than in the United States. She considered having the operation either in Vienna or in Zurich, where her brother's father-in-law was an opthalmologist.[175] Her intention to have eye surgery was opposed by her weakness from the injury and the operation on her arm as well as by her desire to finish her scientific work in Miami. She had agreed to contribute articles on nuclear emulsions[B79] for the series *Methods of Experimental Physics*, edited by Marietta's friend, the physicist C.-S. Wu[A13] (Fig. 26), and her husband, L.C.L. Yuan.

Berta Karlik had offered to let Blau live in her apartment after her return to Vienna; Blau's uncle Hans Golwig was also looking for a place for her to stay.

REMAINING YEARS IN VIENNA

Blau finally returned to Vienna in the spring of 1960. After an absence of twenty-two years, one of the first things she experienced in her native country was the fiftieth anniversary of the Radium Institute; Przibram even mentioned her presence in his talk.[27] She once again became an unpaid worker at the Institute for Radium Research and had to support herself on her pension from the United States, including her medical care, because she did not have health insurance. Her pension of two hundred dollars a month enabled a fairly decent standard of living in Austria at that time, especially since she was said to have lived very modestly.[172]

Those like Marietta Blau who had been expelled by the Nazi regime were not particularly well received upon their return to Austria. In the 1950s and 1960s, Austria had little interest in the return of emigrants, and generally no serious efforts were made to reintegrate successful researchers into scientific life. From 1950 on, those emigrants who had given up their Austrian citizenship had to undergo the same procedure as foreign-born residents in order to establish it again. Blau never did. She also never made the effort to find out whether she was due any restitution and never applied for any.[172]

With the help of a city-council member,[176] she found a centrally located rental apartment that she tastefully decorated with antique furniture.[177] In a letter written shortly after her return,[178] she expressed her joy at seeing the Austrian landscape again and at having a wide variety of musical performances to choose from. In order to celebrate her return to scientific life in Austria, Karl Przibram, together with six other members of the Austrian Academy of Sciences – mainly professors of physics and chemistry at university institutes – proposed her as a corresponding member of the Academy of Sciences.[179] However, the application for membership did not achieve the required majority in the plenary session of the full Academy.[180] At the same time, five of those who had signed this request also nominated her for the Schrödinger Prize. Because the Ministry of Education did not intend to award this prize at all in 1961, it took a follow-up resolution on the part of the Academy's section for mathematics and science[181] reaffirming its position to enable Blau to be recognized the following year. Blau did receive the Schrödinger Prize (thirty thousand Austrian schillings) in 1962 for "the development of the basic photographic method, for the investigation of elementary particles, and in particular for the discovery of disintegration stars together with Dr. Wambacher."[182]

At the Radium Institute, Blau had a small room at her disposal, the *Austragstüberl*, which professors emeriti and retired directors used. As was to be expected, she found herself right in the thick of things as advisor to a high-energy group headed by Brigitte Czapp (later Buschbeck). Walter Thirring,[146] who had been summoned from the University of Bern to accept a professorship in Vienna, had arranged the details. In fact, Thirring was interested in reviving the photographic method in Vienna and therefore had been hoping for Blau's return. He offered her a paid position as leader of this group, but she declined the offer because she feared the obligations might be too demanding for her.[147] Blau would have preferred to earn an extra hundred dollars editing scientific literature, which would have been legally possible for a retiree, but no such opportunity presented itself.

Czapp had finished her studies at the Radium Institute at the end of 1959, before Blau's return to Vienna, and had then spent one year at the University of Bern and at the European Organization for Nuclear Research, CERN, near Geneva preparing herself for the work at the Radium Institute (see p. 138). The group consisted of four students[183] to whom Czapp was to assign dissertation topics, and four women who analyzed photographic plates as well as bubble-chamber photographs. Czapp felt restricted by Blau's critical attitude towards the group, which Blau considered insufficiently provided with equipment.

Blau accepted another female doctoral student, Gerda Haider (later Petkov), whose work[184] she supervised from March 1960 to April 1964 (see p. 143). The plates were scanned by several female students in a large ground-floor laboratory at Boltzmanngasse 5, next to the Radium Institute. Scanning meant the plates were searched for stars, i.e., for divergent tracks of reaction products of proton-proton interactions. Franziska Wagner[177] was particularly skilled in this task but later on gave up studying physics. Another member of this group was Hannelore Eggstain (later Sexl), who started her thesis work at the Radium Institute but later switched to a dissertation at the Institute for Theoretical Physics with Walter Thirring as her advisor. The theoretical consultant of the plate group was Herbert Pietschmann[A14] (Fig. 27; see also p. 140). The photographic plates were sent from CERN to the Vienna Radium Institute as well as to a group in Bern. Each plate was triple scanned, most of the time by three different people, to minimize errors as much as possible. Blau instructed the scanners to send for her immediately when they found certain stars which might indicate hitherto unknown elementary particles. She would then measure these stars herself, despite her poor eyesight. Fig. 28 shows Blau in the laboratory together with her co-workers. In this room Blau sometimes gave talks on high-energy experiments for her collaborators, particularly with respect to the discovery of new elementary particles, before all the Ω^- particle, whose existence had at that time not been definitively proven.[177]

In summer 1961, Blau (Fig. 29) gave talks at the University of Bern as well as at CERN,[165] which gained considerable attention. She traveled to Switzerland together with Eggstain and Haider. Their recollections (see p. 141 and p. 143) express how close

Fig. 27: Herbert Pietschmann in 1966 (Photo courtesy of Herbert Pietschmann).

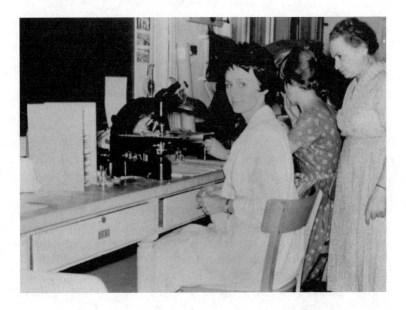

Fig. 28: Marietta Blau and members of the plate group (far left Hannelore Eggstain, later Sexl) in summer 1960 (Archive R. and L. Sexl).

the young women felt to Marietta Blau, beyond their being physically crammed together in the car.

Furthermore, Blau completed the chapters on the application of the photographic method in nuclear physics for *Methods of Experimental Physics*.[B79] Her contribution to this work is one of the finest presentations of this topic in the literature and was also translated into Russian.[B81]

Marietta Blau apparently felt quite alone in Vienna.[177] Among her colleagues from before the war, she met Berta Karlik and Karl Przibram again. She was also in private contact with the younger generation: with the daughter of Helene Aschner, née Pallester, who had attended high school with her, and with Hanne Ellis, née Lauda, who had been Blau's thesis student from 1935 to 1937 (see p. 127). Christine Pfleger, née Lohse, whose mother had been friends with Ludwig Blau's wife Lily in pre-war times, helped Marietta in her everyday difficulties and tried to cheer her up.[23] It was not until 1962 or 1963 that her cataract[1] was operated on.[185]

Fig. 29: Marietta Blau during her stay in Geneva, May 1961 (Archive R. and L. Sexl).

Blau came to the Radium Institute, often on weekends, to be among people and sometimes invited her young co-workers to her comfortable, modest apartment. Now and then she recited poems which her brother Ludwig had written in the style of Joachim Ringelnatz.

After Blau returned to Vienna, her relationship with Berta Karlik did not remain free of tension.[186–188] On the one hand, "the difficult years"[140] had presumably made Marietta Blau bitter, even more so as the victims of National Socialism received no restitution. And then there was the position Stetter received as full professor of physics at the University of Vienna in 1953 and Ortner's as director of the Atom Institute of the Austrian Universities (i.e., the University of Vienna and the Technical University in Vienna) in 1957 and as full professor for nuclear physics at the Technical University in Vienna in 1960. In a letter she expressed her total disappointment at a remark Karlik supposedly made about the past: "... it was only that I was just so bitter for the moment because from various sources I had heard of statements Karlik made about me. And she must know and in fact does know how things used to be at that time, she, Stetter, and Ortner. From the latter two, I would not have expected otherwise, but from her it hurt me deeply."[189] Possibly someone interested in causing discord between the two older women at the Institute relayed this information to her. The fact that, despite her comprehensive scientific achievements, Blau held an unpaid position at the Radium Institute while Karlik served as director may have been the cause of Blau's bitterness.

Years later, Karlik commented on Marietta Blau, stating that "she held herself aloof,"[177] that is to say, that she was unable to gain a reputation on her own. It was for this reason that she attracted so little attention. Blau experienced the differences between the United States and Austria in this regard. In the United States a group makes the effort to integrate individuals who have difficulty with access to others, much more so than in Austria. This is particularly true among intellectuals, and even more so if the individual concerned has the exceptional qualities that Blau had: She was said to be the Pope in the field of the emulsion technique; several of her students became renowned university professors. From her letters to friends in Florida, one learns that in 1962 or 1963 she thought of returning to the United States, in particular to New York.[185] In her mind, happiness was to be found elsewhere. She had dreamed of Austria when she lived in the United States, but now it was the other way round.

In Austria, Blau remained at the periphery. The reason for this may have been a certain distrust of people she met here. If this were the case, she may have shared this attitude with many emigrants who could not forget the events of 1938. Collective persecution had yielded a collective mistrust of people, with little differentiation as to their political standpoint during the Nazi era. This may explain why Blau never spoke about her emigration with her students, even with those to whom she was close personally or why she only spoke to Arnold Perlmutter and Leopold Halpern about her disappointment with her position at the Vienna Radium Institute. Individuals working at the Institute at the beginning of the 1960s describe her reserved behavior as follows: "She always seemed completely withdrawn. When one spoke to her, she opened up extremely cautiously, and then closed up again."[190] "I met her on the staircase once in a while but never had a conversation with her. She was very shy, almost unsociable."[191] "She was nice and friendly."[192] Despite being distant with many of her fellow workers, it was her way to express her love abundantly to those who showed signs of liking her: "I loved her a lot. She was like a surrogate mother to me."[193] "We had a lovely relationship."[187] Seemingly Blau dealt skeptically and even shyly towards strangers with whom she shared neither scientific nor personal interests but demonstrated an unwavering certainty in scientific dialogue, a remarkable willingness to guide scientific projects, and last but not least, impressive perseverance and endurance in her affection once she had befriended someone.

Apart from a relationship with one of her brother's friends in pre-war Vienna, Marietta Blau cultivated the friendship of women more than that of men. But most likely she never committed herself completely to another person.[11]

Among the few who came to see her was Leopold Halpern, professor at Florida State University in Tallahassee, who had first met Blau when she was working at Brookhaven National Laboratory.

After Blau's last doctoral student, Gerda Haider, had graduated at the end of 1964, Blau found herself unable to walk long distances or to get on trams or buses because of her health and therefore decided to quit her activities at the Radium Institute. She also abandoned her dream to visit the United States again. Fully resigned, she wrote to Arnold Perlmutter that she was afraid she would never see the United States again.[194] She spent much time by herself in her apartment, where loneliness may have contributed to her poor state of health.

At the end of February 1966, she went to the Lainz hospital for the first time. In the anamnesis, her year-long heart problems were specified as sclerosis of the coronary vessels of the heart.

♦ The symptoms of this heart condition, also called *angina pectoris*, are described as follows: "Acute insufficiency of the coronary vessels of the heart with suddenly onsetting pain lasting for seconds to minutes, which radiates to the left (right) shoulder-hand-arm-region or throat-lower jaw-region, respectively, often feelings of belt-like constriction around the chest with spells of suffocation and asthma reaching to feelings of destruction and mortal terror."[195]

Interpreting a heart condition as a psychosomatic symptom, this illness is assumed to express that one does not give enough attention to one's own needs but instead dedicates oneself to some task entirely, and completely neglects taking good care of oneself.[196] ♦

According to Blau's hospital report: "Since 1962 strong stabbing pain in the region of the heart, radiating into the left arm ... For two years, spells of shortness of breath during nights." In addition: "Walking unstably in January 1966 and half a year ago sudden speech disorder with weakness in the right hand. The speech disorder receded completely, but a certain weakness in the right hand remained (patient cannot write, primarily)."[197] After two weeks, she went home after some improvement in her condition.

In 1967 Marietta Blau received the Science Prize of the City of Vienna[198] and a plaque from the Paris Radium Institute on the occasion of Marie Curie's hundredth birthday.[172] By the end of September 1967, she again required treatment in the Lainz hospital for pain in her abdomen and problems with her liver and bladder.[197] Again, she was sent home when she showed some improvement after a stay of seventeen days. When she entered the hospital, Blau had named an acquaintance[197] as the person to contact in case of emergency instead of anyone from the Radium Institute.

The end of March 1969 was the fiftieth anniversary of Blau's Ph.D. graduation. Berta Karlik, as the director of the Radium Institute, applied for renewal of Blau's doctoral diploma (an honor the University of Vienna pays to alumni of high merit), "who by systematic investigations had essentially furthered the method of nuclear emulsions and by its use had achieved valuable research results in high-energy physics."[199] Blau received a letter from the dean of the faculty[200] inviting her to his office to personally receive this honor. She replied with delight but evasively,[201] since she planned to visit her brother Otto in Lugano. Her trip was prevented by new health problems that resulted in another stay at the Lainz hospital from April 4 to 16, 1969. As the person to contact, Blau had named only the caretaker of her apartment house. In addition to the symptoms noticed during the two previous hospital stays, a shadow was found on her right lung. After remaining in the hospital for twelve days, she returned home; the discharge report states laconically: "At the patient's request, further examinations have been cancelled and the patient has been dismissed to home care."[197] The visit to her brother took place in May 1969. By the middle of July 1969, she wrote the dean again, this time after having received the Golden Doctoral Diploma from Berta Karlik. Blau's state of health had not permitted her to attend the presentation at the dean's office, but she promised to visit at a later time.[202]

Also in 1969, she received letters of recognition from the mayor and the city councillor for culture and public education.[172] Because of her gall bladder ailment, she had to enter the Lainz hospital for treatment again at the end of July 1969; the diagnosis was a tumor in her right lung and suspicion of another in her pelvis.[197] The lung tumor had become apparent in the x-ray examination four months earlier but had not been diagnosed conclusively because she had refused further examinations. Now she had cancer, as other physicists of the Radium Institute had had before her: Hönigschmid (died 1945), Schweidler (died 1948), and her collaborator Wambacher (died 1950). Marietta Blau's tuberculosis, the handling of open radioactive preparations, and decades of smoking were certainly factors in her present illness. Again she was dismissed from hospital after only fifteen days. Since she had to pay for treatments herself, she seems to have tried to keep her hospital stays as short as possible. Six weeks later, Blau's status became critical: Rapidly losing weight, she had to be readmitted for treatment; her brother Otto came to Vienna to accompany her. (Her brother Ludwig had died in 1966, also from cancer.[129]) The progressing lung cancer metastasized; Blau suffered from shortness of breath and pain radiating from her heart.[197]

After four months in the hospital, Blau died at noon on January 27, 1970, from lung edema and circulatory collapse.[197] At her own request, her body was cremated privately and the urn buried in her father's grave (Fig. 30) at the Vienna Central Cemetery.[203] In the obituary notice, her brother Otto characterized her: "Her life was dedicated to science and filled with kindness and charity."[203]

In an announcement of Blau's death in the almanac of the Austrian Academy of Sciences, Karlik wrote: "The Institute regrets the death of an excellent member of long standing, Prof. Dr. Marietta Blau, who passed away in a hospital in Vienna after a long illness on January 27, 1970. Marietta Blau pioneered the method of nuclear emulsions and applied this method with great success, in particular in high-energy physics. The Austrian Academy of Sciences honored her with the Schrödinger Prize in 1962 ... An obituary will be published in *Acta Physica Austriaca*."[204] However, the notice never appeared because Karlik herself fell ill for some months in 1970. No scientific journal ever published an obituary for Marietta Blau.

Fig. 30: Grave of Marietta Blau and her father, Vienna Central Cemetery (Photo by Brigitte Strohmaier, 2000).

NOTES

[A1] Stefan Meyer (born April 27, 1872, Vienna; died December 29, 1949, Bad Ischl) graduated from the state high school in Horn, Lower Austria, in 1891 and started studying physics, chemistry, and mathematics at the University of Vienna one year later. He worked on his dissertation with Franz S. Exner on the location of the potential differences in drip electrodes and in the capillary electrometer and received his Ph.D. in 1896. Meyer started his scientific career as an assistant to Ludwig Boltzmann. He measured the magnetic permeabilities of elements (together with Gustav Jäger) and discovered the deflexibility of radium and polonium rays in magnetic fields (together with Egon von Schweidler). In 1900 he acquired his university teaching certification (*venia legendi*) with his work on radium and polonium rays (according to later terminology: β- and α-rays). From 1902 to 1911, he was associate professor for acoustics at the Vienna School of Music, in 1906/07 provisional director of the Institute for Theoretical Physics. In 1909 he started working on discoloration and luminescence of solids due to radioactive irradiation. He was put in charge of planning and supervising the construction and equipment of the Radium Institute due to his experience in work on radioactivity. However, in 1910 he was appointed First Assistant of the Institute, although in fact he functioned as its head (Franz S. Exner was the director in name). Only in 1920 did Meyer officially become its director. As a sign of international recognition, he became secretary of the International Radium Standard Commission in 1910. Further steps in his career were: 1910, award of the Franz Joseph order; 1911, associate professor; 1913, Lieben Prize; 1915, title and privileges of full professor; 1916, publication of *Radioaktivität* with Egon von Schweidler; 1920, position of full professor; 1921, corresponding member of the Academy of Sciences in Vienna; 1927, second edition of *Radioaktivität*, 1932, full member of the Academy of Sciences in Vienna. In 1937, he was elected president of the International Radium Standard Commission. As a descendant of a Jewish family, he was dismissed from his position at the Radium Institute by the Nazis in 1938. He survived the war in seclusion in Bad Ischl, protected thanks to his daughter's courage and diplomatic skills. After Austria's liberation in 1945, he was reappointed as honorary professor and retired in 1947.

[A2] Karl Przibram (born December 21, 1878, Vienna; died August 10, 1973, Vienna), obtained his *Matura* at the Akademisches Gymnasium in Vienna in 1896. He studied physics, chemistry, and mathematics at the University of Vienna; his teachers were Franz S. Exner and Ludwig Boltzmann. He worked on his doctorate in Graz titled *Contributions to the Understanding of the Differing Performance of Anodes and Cathodes at Electric Discharge* with Leopold Pfaundler.[205] He received his Ph.D. in 1901. After a year (1902/03) at the Cavendish Laboratory in Cambridge with J.J. Thomson, he did experimental work at Boltzmann's institute in Vienna. With work on brush discharge, he acquired university teaching certification in 1905. In 1912 he came to the Institut für Radiumforschung and became assistant in 1920. His studies concentrated on the discoloration and luminescence of solids caused by radioactive radiation (radiophoto luminescence, radiothermo luminescence, radiophoto fluorescence). In 1914 he was awarded the Haitinger Prize of the Academy of Sciences in Vienna, in 1929 the Lieben Prize. In 1927 he was promoted to associate professor at the University of Vienna. Przibram also had impressive artistic talent.

After Austria's annexation by Nazi Germany in 1938, Przibram was dismissed from the university because of his Jewish origin and emigrated to Belgium in 1939, where he survived the years of German occupation underground. He wrote reports on technical subjects for the Union Minière du Haut Katanga between 1940 and 1946. In 1946 he received an appointment at the University of Vienna and returned there in 1947 as full professor and director of the Second Physics Institute, which he completely restructured. He was a pioneer in the field of influencing solids by radiation (e.g., discoloration of alkali halides). In 1947 he was elected corresponding member and in 1950 full member of the Austrian Academy of Sciences. After becoming professor emeritus in 1951, he returned to the Radium Institute for further research, which is presented in his book[206] *Verfärbung und Lumineszenz* (Discoloration and Luminescence).

In 1963 he received the Schrödinger Prize of the Austrian Academy of Sciences.

[A3] Berta Karlik (born January 24, 1904, Mauer near Vienna; died February 4, 1990, Vienna) attended the Reformrealgymnasium for girls in Vienna-Hietzing and studied physics and mathematics at the University of Vienna. In her dissertation at the Radium Institute, she investigated the scintillation method, which was predominantly used for particle detection there at the time. She received her Ph.D. in 1928, followed by certification for the teaching profession, and spent the year of probation as a highschool teacher,[35] followed by one year of unpaid research work at the Radium Institute.

From 1933, Karlik held the position of research assistant at the Vienna Radium Institute. Together with Elisabeth Rona, she won the 1933 Haitinger Prize of the Academy of Sciences. In 1937 her *Habilitation* work, *Detection Limits of the Heavier Noble Gases in Helium,* was approved, bestowing on her the right of teaching at the university; in 1940 she became a university assistant, in 1942 a university lecturer with pay. Together with Traude Bernert, in 1943 she discovered the element with the atomic number eighty-five in natural radioactive-decay series. This element was later called astatine; Karlik and Bernert had initially proposed the name viennium. After Austria was liberated in 1945, Karlik was installed as interim director of the Radium Institute; at the same time, she reestablished the Austrian Association of University Women (Verband der Akademikerinnen Österreichs). In 1946 the title of associate professor was conferred on her. (In Austria, being granted the title of associate professor is a promotion lesser than being appointed associate professor.) In 1947 Karlik became the permanent director of the Radium Institute and was awarded the Haitinger Prize for physics for her work on element eighty-five. Further steps in her university career were being appointed associate professor in 1950 and becoming the first female to reach the rank of full professor at the University of Vienna in 1956. The Austrian Academy of Sciences elected her corresponding member in 1954 and full member in 1973. As professor emerita (1974), she worked on historical subjects, culminating in the publication of the book *Franz S. Exner und sein Kreis* (Franz S. Exner and His Circle),[207] together with Erich Schmid in 1982.

[A4] Hans Pettersson (born August 26, 1888, Forshälla, Sweden; died January 25, 1966, Göteborg, Sweden) studied physics at the University of Uppsala and spent a year (1911/12) with William Ramsey at the University College in London. He obtained his Ph.D. at the University of Göteborg in 1914, where he then lectured on oceanography. In connection with the determination of the radium content of deep-sea sediments from the *Challenger* expedition, he started collaborating with the Vienna Radium Institute and the Second Physics Institute of the University of Vienna in 1922; this cooperation lasted until 1936. From 1922 to 1930, he spent several months per year in Vienna. There he initiated nuclear-reaction studies and stimulated the development of various working methods in nuclear physics. In 1930 he became full professor of oceanography in Göteborg; due to his initiative, the new Oceanographic Institute was founded. He studied the radioactivity of sea water, as well as other areas of oceanography, and organized and directed the Swedish deep-sea expedition *Albatros* in 1947/48. Besides sediment samples of up to sixteen meters in length, he also investigated cosmic dust reaching the earth.

[A5] Gerhard Kirsch (born June 21, 1890, Vienna; died September 15, 1956, Bischofshofen, Salzburg) studied chemistry and then physics at the University of Vienna and earned his Ph.D. in 1920 after interrupting his studies for military service in World War I. He became a university assistant in 1921, a lecturer in 1925, and an associate professor in 1932. His work concerned radioactive-decay series and dating of minerals. In collaboration with Hans Pettersson, he worked on α-induced nuclear reactions and took part in the Swedish hydrographic expedition in the Baltic Sea. From 1928 on, he mainly worked in the interdisciplinary field of geology and radioactivity. After Austria's annexation by Germany in 1938, he directed the Third Physics Institute of the University of Vienna; in 1941 he became head of the First Physics Institute at the University of Vienna and of the Research Institute Gastein, where he worked mainly on the analysis of radioactive spas. In 1945 he was dismissed from the University because of his affiliation (since 1933) with the Nazi party and was forced to retire in 1947.

[A6] Viktor Hess (born June 27, 1883, Waldstein Castle near Graz; died December 17, 1964, Mt. Vernon, New York) studied physics at the University of Graz and obtained his Ph.D. (*sub auspiciis imperatoris*) in 1906. Afterwards, he worked at the University of Vienna with Franz Exner. In 1910 his *Habilitation* work, *Absolute Determination of the Atmosphere's Contents of Radium Progeny*, was approved; the same year he became First Assistant at the Radium Institute and remained in this position until 1920. He measured the ionization of the atmosphere during numerous balloon flights in 1911/12. Through this research, he discovered cosmic rays, for which he was granted the Nobel Prize in Physics in 1936 (together with C.D. Anderson, who had discovered the positron). In 1925 he was appointed full professor at the University of Graz and in 1931 at the University of Innsbruck. He installed a small observatory on Hafelekar, north of Innsbruck, where the intensity of cosmic rays was constantly registered. In 1937 he returned to Graz and was dismissed in 1938 because of his opposition to the Nazis. He was permitted to emigrate to the United States after his Nobel Prize was seized. He became professor at Fordham University in New York; from 1956, he was professor emeritus. In 1959 he was awarded the Österreichisches Ehrenzeichen für Kunst und Wissenschaft but never returned permanently to Austria.

[A7] Ellen Gleditsch (born December 29, 1879, Mandal, Norway; died June 5, 1968, Oslo, Norway) attended a private high school but as a woman was not allowed to take the necessary exams to enter the university. As a pharmacy assistant, she obtained a non-academic degree in chemistry and pharmacology, which entitled her to study at the University of Oslo. In 1905/06 she passed the exams qualifying her for regular university studies.

Gleditsch went to Paris in 1907, where she worked for Marie Curie at the Paris Radium Institute; at the same time, she studied at the Sorbonne and completed her Ph.D. in 1912. Gleditsch went to Yale on a fellowship in 1913. With a precision measurement of radium half-life, she earned a reputation as the leading expert in the preparation of radioactive substances from minerals. Back in Oslo, she pioneered the uranium-series method for dating minerals and became a university lecturer for radiochemistry in 1916. She published several books on inorganic and radio chemistry and returned to the Institut Curie in Paris several times. She was also a member of the Academy of Sciences in Oslo. She was president of both the International and the Norwegian Federation of University Women in the 1920s, until she was appointed full professor for inorganic chemistry at the University of Oslo in 1929.

Gleditsch was actively involved in the resistance against Nazi Germany. After the war, she was active in the newly founded UNESCO. She retired at the beginning of 1946 but continued her scientific and political activities. She received many honors and prizes and is given more credit for her humanitarian efforts than for her impressive scientific achievements.

[A8] Hans Thirring (born March 23, 1888, Vienna; died March 22, 1976, Vienna) graduated from the Sophiengymnasium in 1907 and began university studies in mathematics, physics, and sports the same year. Fascinated by Fritz Hasenöhrl's teaching, Thirring decided to make theoretical physics the focus of his studies and wrote a dissertation, *On Some Thermodynamical Relationships near the Critical and the Triple Point*. In 1910 he took the examination for teaching physical education, two years later that for physics and mathematics. Also in 1910 he became Hasenöhrl's assistant; the following year he obtained his Ph.D. He received his university teaching certification with a work on *The Application of Canal Rays on Gas Analysis*. After Hasenöhrl's death in World War I, Thirring taught at the University and at the Technical University in Vienna, where he conducted successful work in relativity (the Thirring-Lense effect). In 1921 Thirring became associate professor and head of the Institute for Theoretical Physics of the University of Vienna, in 1927 full professor. Among other innovations, he developed selenium cells and invented the *Flattermantel* for hover skiing. Because of his antifascist and pacifist convictions, he was dismissed on December 1, 1938. He held industry positions with Elin and Siemens-

Halske and wrote the book *Homo Sapiens – Psychologie der menschlichen Beziehungen* (Psychology of Human Relationships). From 1945 to 1958, he was again full professor at the Institute for Theoretical Physics and worked actively for peace in the world (the Pugwash movement, Thirring plan).

[A9] Georg Stetter (born December 23, 1895, Vienna; died July 14, 1988, Vienna) obtained his *Matura* in 1914 and started studying mechanical engineering and electrotechnology at the Technical University in Vienna. His military service in a telegraphy regiment in World War I is said to have laid the basis for his interest in electromagnetic waves. He continued his studies at the University of Vienna, earned his Ph.D. in 1922 and became an assistant at the Second Physics Institute. Because of its close connection to the Radium Institute, he also explored nuclear physics. He pioneered the usage of valves for the measurement of particle energies in nuclear reactions which led to today's important field of nuclear electronics. He obtained his university teaching certification on the basis of this work in 1928. With colleagues, he precisely measured the energy released in uranium fission. In 1938 he was appointed professor and became head of the Second Physics Institute. He designed a nuclear reactor and registered a patent. In 1943 he became head of the Four-Year-Plan Institute for Neutron Physics. After he was removed from the university in 1945, he designed a dust monitor for coal mines. In 1953 he was again appointed professor and director of the First Physics Institute. In 1962 he was elected a full member of the Austrian Academy of Sciences and founded its commission on clean air. He is viewed as a major influence on the development of physics in Austria.

[A10] Gustav Ortner (born July 31, 1900, Haus/Ennstal, Styria; died November 24, 1984, Afling, Tyrol) studied mathematics and physics at the University of Vienna. His research began at the Vienna Radium Institute at a time when atomic and nuclear physics were developing rapidly. With his friend Georg Stetter, he developed proportional counters there in the 1920s and 1930s; these devices allowed spectacular progress in the detection of ionizing radiation and contributed substantially to the discovery of the neutron and of nuclear fission. In 1939 Ortner became associate professor at the University of Vienna and head of the Vienna Radium Institute. During the war, he was part of a group working on the development of a nuclear reactor. In 1945 he was detained in Moscow for several months. After his return to Vienna, he was dismissed because of his NSDAP affiliation. In 1950 he became professor at the University of Cairo in Egypt. He returned to Vienna in 1955, when a program for the peaceful use of nuclear physics also led to planning for research reactors in Austria. He became the head of the Atom Institute of the Austrian Universities, which he directed with distinction. In 1960 Ortner was hired as professor of nuclear physics at the Technical University in Vienna.

[A11] Cecil Frank Powell (born December 5, 1903, Tonbridge, Great Britain; died August 9, 1969, Milan, Italy) studied at Cambridge University with E. Rutherford and C.T.R. Wilson. He obtained his Ph.D. in 1927 and was then assistant to A.M. Tyndall at Bristol University, where he investigated condensation phenomena and took part in the expedition that explored the eruption of the Montserrat volcano. In 1938 he turned to nuclear physics, particularly to the photographic method. He became professor in 1948 and Fellow of the Royal Society in 1949. He was awarded the Nobel Prize in Physics in 1950. Powell was president of the Continuing Committee of the Pugwash Conferences, of the Scientific Policy Committee at CERN, Geneva, and of the World Federation of Scientific Workers.

[A12] Arnold Perlmutter (born November 4, 1928, Brooklyn, New York) studied physics at New York University from 1949 to 1955, when he received his Ph.D. From 1956 to 1968 he was university assistant and lecturer for elementary-particle physics at the University of Miami in Florida. In 1968 he was promoted to full professor. Starting in 1965, he was secretary-general of the Center for Theoretical Studies. His fields of research were photoconductivity of phosphors, nuclear emulsions, elementary particles, potential scattering computations, electromagnetic radiation, polarized neutron scattering, and global energy. He has been professor emeritus since 2001.

A13 Chien-Shiung Wu (born May 31, 1912, Shanghai; died February 16, 1997, New York City) began her studies in China, came to the U.S.A. in 1936, and received her Ph.D. in physics at the University of California, Berkeley in 1940. She taught at Smith College and Princeton University before coming to Columbia University in 1944. During World War II, she took part in the Manhattan Project as a nuclear physicist. In 1956 when Chen Ning Yang and Tsung-Dao Lee postulated that parity was not preserved in weak interaction on the basis of an analysis of existing data in particle physics, Wu was the first to carry out an experiment together with a group of physicists from Columbia University and the National Bureau of Standards that confirmed their conclusions. Together with Yang and Lee, she was nominated for the Nobel Prize in Physics in 1957, which was awarded to Yang and Lee. Wu became full professor in 1958 and the first Pupin Professor of Physics in 1973; she also held honorary professorships at several Chinese universities. Wu co-authored the book *Beta Decay*.[208] She was the first woman to be president of the American Physical Society. She retired in 1980.

A14 Herbert Pietschmann (born August 9, 1936, Vienna) finished his studies in physics at the University of Vienna in 1960 and completed his Ph.D. (*sub auspiciis praesidentis*) in 1961. In 1964/65, he was a visiting scientist at the University of Virginia. After his return to Vienna, he was certified to teach theoretical physics at the University of Vienna in 1966 and was university professor at the Technical University in Göteborg and then guest professor at the University of Bonn. He returned permanently to Vienna in 1968 and was promoted to full professor in 1971. He worked in the field of theoretical particle physics, in particular on weak interaction and quantum field theory within unified field theories. He also published numerous books on philosophical topics. He has been professor emeritus since October 2004.

Abbreviations used:
MIR   *Mitteilungen des Instituts für Radiumforschung*
Archive ÖAW   Archive of the Austrian Academy of Sciences (Österreichische Akademie der Wissenschaften), Vienna.
Sondersammlung Zentralbibl. Phys.   Special collection of the Österreichische Zentralbibliothek für Physik, Vienna.

[1] Wolfgang L. Reiter, "Marietta Blau," *Vertriebene Vernunft*, ed. F. Stadler (Vienna and Munich: Verlag für Jugend und Volk, 1987), 720.
[2] Friedrich Katscher, "Jüdische Naturwissenschaftler und Techniker Österreichs – Was sie der Welt schenkten," *Österreichisch-jüdisches Geistes- und Kulturleben*, ed. Liga der Freunde des Judentums (Vienna: Literas-Universitätsverlag, 1988), 80.
[3] Brigitte Bischof, "Marietta Blau (1894–1970)," *Wissenschaft und Forschung in Österreich*, ed. Gerhard Heindl (Frankfurt/Main: Peter Lang – Europäischer Verlag der Wissenschaften, 2000), 147.
[4] Leopold Halpern, "Marietta Blau (1894–1970)," *Women in Chemistry and Physics*, eds. L.S. Grinstein, R.K. Rose, M.H. Rafailovich (Westport, Connecticut and London: Greenwood Press, 1993), 57.
[5] Leopold Halpern, "Marietta Blau – Discoverer of the Cosmic Ray 'Stars'," *A Devotion to Their Science*, eds. M.F. Rayner-Canham and G.W. Rayner-Canham (Montreal & Kingston, London, Buffalo: McGill Queen's University Press, 1997), 196.
[6] Peter L. Galison, "Nuclear Emulsions – The Anxiety of the Experimenter," *Image and Logic – A Material Culture of Microphysics* (Chicago: University of Chicago Press, 1997), 143.
[7] Peter L. Galison, "Marietta Blau – Between Nazis and Nuclei," *Physics Today* 50 (1997) 42.
[8] N. Byers, "Contributions of 20th Century Women to Physics," http://www.physics.ucla.edu/~cwp/phase 2/Blau,_Marietta@843727247.html.
[9] Addresses of the Blau family's dwelling places:
1891: Wien 2, Praterstraße 21; 1893: Wien 3, Obere Weißgerberstraße 7; 1894: Wien 2, Schmelzgasse 6; 1900: Wien 3, Hintere Zollamtsstraße 13; 1911: Wien 2, Taborstraße 20; 1913: Wien 1, Morzinplatz 6; 1933: Wien 19, Grinzingerstraße 93 (Florentine, Otto, Marietta) and Wien 19, Grinzingerstraße 86 (Ludwig with wife and daughter).

[10] Marsha L. Rozenblit, *Die Juden Wiens 1867–1914* (Vienna: Böhlau, 1989), 80–98.

[11] Eva Connors (Marietta Blau's niece), telephone communication to Brigitte Strohmaier, August 16, 2000.

[12] *Von hier nahm die höhere Mädchenbildung in Österreich ihren Ausgang* (Here began women's higher education in Austria), inscription on the building at Rahlgasse 4.

[13] Joseph Aschner, son of Helene Aschner-Pallester, personal communication to Robert Rosner, January 11, 2000.

[14] Archive of the Bundesgymnasium and Bundesrealgymnasium at Rahlgasse 4, Vienna.

[15] Margarete Meschkan (*Matura* 1928, Rahlgasse), personal communication to Brigitte Strohmaier, February 28, 2000.

[16] Before 1911, women teachers in Vienna were not allowed to marry.

[17] Marietta Blau, letter to Eva Blau, her niece, Mexico City, August 19, 1942, courtesy of Eva Connors.

[18] Archive of the University of Vienna, communication to Brigitte Strohmaier, January 27, 2000.

[19] Founding document, Institute for Radium Research, *Almanach Kaiserl. Akad. Wiss.* 61 (1911) 212.

[20] Stefan Meyer, "Die Vorgeschichte der Gründung und das erste Jahrzehnt des Institutes für Radiumforschung," *Sitzungsber. Österr. Akad. Wiss., Math. Naturwiss. Kl. IIa* 159 (1950) 1.

[21] Among the women who obtained their Ph.D. in physics at the same time as Blau were Maria Anna Schirmann, one of Marietta's former classmates, and Elisabeth Bormann, an aunt of the editor of Ariadne Press, Jorun Johns.

[22] Death announcement of Markus Blau, *Neue Freie Presse*, December 20, 1919.

[23] Eva Connors, letter to Brigitte Strohmaier, September 19, 2000.

[24] Registration office of the Jewish Community in Vienna.

[25] Marietta Blau, postcard to Stefan Meyer, Berlin-Schöneberg, July(?) 16, 1921, Archive Radiumforschung, Archive ÖAW.

[26] Wolfgang L. Reiter, "Stefan Meyer – Pioneer of Radioactivity," *Physics in Perspective* 3 (2001) 106.

[27] Karl Przibram, in: "Das 50jährige Bestandsjubiläum des Institutes für Radiumforschung," *MIR 550; Sitzungsber. Österr. Akad. Wiss., Math. Naturwiss. Kl. II* 170 (1960) 239.

[28] Engelbert Dollfuss, "'Fremdländer'-Frage in der Wiener Universität," *Reichspost*, September 24, 1920.

[29] Hans Pettersson, letter to his sister, 1926, cited by Agnes Rodhe in a letter to Robert Rosner, Lund, Sweden, April 12, 2000.

[30] Karl Przibram, "1920–1938," *Sitzungsber. Österr. Akad. Wiss., Math. Naturwiss. Kl. IIa* 159 (1950) 27.

[31] Berta Karlik, communication to Auguste Dick, March 5, 1984, Sondersammlung Zentralbibl. Phys.

[32] Wilhelm Michl, "Über die Photographie der Bahnen einzelner α-Teilchen," *Sitzungsber. Kaiserl. Akad. Wiss. Wien, Math. Naturwiss. Kl. IIa* 121 (1912) 1431; Wilhelm Michl, "Zur photographischen Wirkung der α-Teilchen," *Sitzungsber. Kaiserl. Akad. Wiss. Wien, Math. Naturwiss. Kl. IIa* 123 (1914) 1955.

[33] Roger H. Stuewer, "Artificial Disintegration and the Cambridge-Vienna Controversy," *Observation, Experiment and Hypothesis in Modern Physical Science*, eds. Peter Achinstein and Owen Hannaway (Cambridge, Mass., London: MIT Press, 1985), 239.

[34] Wolfgang L. Reiter, "Elisabeth Rona," *Vertriebene Vernunft*, ed. F. Stadler (Vienna and Munich: Verlag für Jugend und Volk, 1987), 718.

[35] At the Realgymnasium in Albertgasse in the eighth district.

[36] Elise Richter graduated in Romance languages and was the first woman to achieve *Habilitation* (right of teaching) at the University of Vienna.

[37] Marianne Hainisch (1839–1936) demanded founding high schools (*Realgymnasien*) for girls and access to university studies for women in 1870. In 1902, she founded the League of Austrian Women's Associations (1914: ninety associations), whose presidency she held until 1918. She was the mother of Michael Hainisch (1858–1940), Austrian social and economic politician and first president of the Austrian Republic (1920–1928).

[38] *Mitteilungen des Verbands der Akademikerinnen Österreichs* 66, Special issue 3A (1997).

[39] *Almanach Akad. Wiss. Wien* 83 (1933) 295.

[40] Herta Leng, "Adsorptionsversuche an Gläsern und Filtersubstanzen nach der Methode der radioaktiven Indikatoren," *MIR* 195; *Sitzungsber. Akad. Wiss. Wien, Math. Naturwiss. Kl. IIa* 136 (1927) 19.

[41] Elizabeth Rona, "Laboratory Contamination in the Early Period of Radiation Research," *Health Physics* 37 (1979) 723.

[42] Stefan Meyer and Egon Schweidler, *Radioaktivität* (Leipzig and Berlin: Verlag B.G. Teubner, 1927), 262.

[43] The fact that – contrary to Meyer's description – Blau suffered these injuries to her right hand, suggests – beside other indications – that she was left-handed; she wrote with her right hand, though.

[44] Stefan Meyer, "Otto Hönigschmid (Obituary)," *Anzeiger Akad. Wiss. Wien* 82 (1945) 23.

[45] Hertha Wambacher, Curriculum vitae, Sondersammlung Zentralbibl. Phys.

[46] Hertha Wambacher, "Untersuchung der photographischen Wirkung radioaktiver Strahlungen auf mit Chromsäure und Pinakryptolgelb vorbehandelte Filme und Platten," *MIR* 274; *Sitzungsber. Akad. Wiss. Wien, Math. Naturwiss. Kl. IIa* 140 (1931) 271.

[47] Marietta Blau, autobiographical sketch (transmitted by Leopold Halpern, who had received it from Marietta Blau's brother Otto Blau in 1977), Sondersammlung Zentralbibl. Phys.

[48] Marietta Blau and Herta Leng, postcard to Stefan Meyer, Grimmenstein, January 29, 1929, Archive Radiumforschung, Archive ÖAW.

[49] Herta Leng, "Zur Frage der photographischen Wirksamkeit sonnenbestrahlter Metalle," *MIR* 262a (1930).

[50] Robert W. Pohl (1884–1976), professor of physics at Göttingen.

[51] John Eggert (1891–1973), head of the scientific central laboratory of the photographic department of Agfa.

[52] Marietta Blau, letter to Stefan Meyer, Leipzig, October 4, 1932, Archive Radiumforschung, Archive ÖAW.

[53] Walter Heitler (1904–1981), German physicist who worked at Göttingen University 1929–1933, emigrated to England, where he did research in Bristol, later in Dublin. After the war, he went to Zurich, Switzerland.

[54] Eduard G. Steinke (1899–1963), 1937–1945, professor in Freiburg; 1950–1956, at the University of Santa Fe, Argentina; 1956, return to Germany; 1957–1959, at the University of Stuttgart; 1960, Center for Nuclear Research, Karlsruhe.

[55] James Franck (1882–1964), Nobel Prize in Physics in 1925 together with Gustav Hertz; 1920–1933, in Göttingen; 1934, Baltimore University; 1938, Chicago (photosynthesis, Manhattan Project); initiated the Franck Report, which demanded in June 1945 that the atomic bomb not be used.

[56] Marietta Blau, letter to Berta Karlik, Göttingen, October 22, 1932, Archive Radiumforschung, Archive ÖAW.
[57] Arnold Eucken (1884–1950), professor of physical chemistry, University of Göttingen.
[58] Adolf Gustav Smekal (1895–1959), professor of physics at the University of Vienna, later Halle, Germany, and Graz.
[59] Marietta Blau, letter to Berta Karlik, Göttingen, November 13, 1932, Archive Radiumforschung, Archive ÖAW.
[60] Eugen Rabinowitsch (1903–1973), physical chemist; born in St. Petersburg; 1926, Ph. D. at University of Berlin; research associate Göttingen, London, and MIT; Manhattan Project; from 1947, University Illinois.
[61] Andreas von Antropoff (1878–1956), professor of chemistry, University of Bonn.
[62] Marietta Blau, letter to Stefan Meyer, Göttingen, December 12, 1932, Archive Radiumforschung, Archive ÖAW.
[63] Marietta Blau, letter to Stefan Meyer, Göttingen, February 8, 1933, Archive Radiumforschung, Archive ÖAW.
[64] Stefan Meyer, letter to Marietta Blau, Vienna, February 14, 1933, Archive Radiumforschung, Archive ÖAW.
[65] Marietta Blau, letter to Stefan Meyer, Göttingen, February 18, 1933, Archive Radiumforschung, Archive ÖAW.
[66] Salomon Rosenblum (1896–1959), physicist, one of Marie Curie's closest collaborators.
[67] Marietta Blau, letter to Stefan Meyer, Paris, April 29, 1933, Archive Radiumforschung, Archive ÖAW.
[68] Karl Przibram at that time was deputy director of the Radium Institute.
[69] Marietta Blau, letter to Stefan Meyer, Paris, June 10, 1933, Archive Radiumforschung, Archive ÖAW.
[70] Marietta Blau, letter to Stefan Meyer, Paris, June 29, 1933, Archive Radiumforschung, Archive ÖAW.
[71] Marietta Blau, letter to Stefan Meyer, Paris, July 19, 1933, Archive Radiumforschung, Archive ÖAW.
[72] Elvira Steppan, "Das Problem der Zertrümmerung von Aluminium behandelt mit der photographischen Methode," *MIR* 370; *Sitzungsber. Akad. Wiss. Wien, Math. Naturwiss. Kl. IIa* 144 (1935) 455.
[73] Stefanie Zila, "Beiträge zum Ausbau der photographischen Methode für Untersuchungen mit Protonenstrahlen," *MIR* 386; *Sitzungsber. Akad. Wiss. Wien, Math. Naturwiss. Kl. IIa* 145 (1936) 503.

[74] Hanne Lauda, "Über das Abklingen des latenten Bildes auf der photographischen Platte," *MIR* 390; *Sitzungsber. Akad. Wiss. Wien, Math. Naturwiss. Kl. IIa* 145 (1936) 1.

[75] Johanna Riedl, "Über die Gruppenstruktur der Rückstoßprotonen von α-Teilchen," *MIR* 416; *Sitzungsber. Akad. Wiss. Wien, Math. Naturwiss. Kl. IIa* 147 (1938) 181.

[76] Otto Merhaut, *Das Problem der Resonanzeindringung von α-Teilchen in den Aluminiumkern, behandelt mit der photographischen Methode*, Dissertation, University of Vienna (1938).

[77] *Almanach Akad. Wiss. Wien* 87 (1937) 351.

[78] Robert Rosner, "Der Ignaz-Lieben-Preis," *Chemie* 4 (1997) 30.

[79] Protocol of the plenary session of the Austrian Academy of Sciences, Vienna, April 2, 2004.

[80] Rudolf Steinmaurer, "Erinnerungen an V.F. Hess, den Entdecker der kosmischen Strahlung, und an die ersten Jahre des Betriebes des Hafelekar-Labors," *Early History of Cosmic Ray Studies*, eds. Y. Sekido and H. Elliot (Dordrecht, Boston, London: D. Reidel, 1985), 28.

[81] Marietta Blau, letter to Stefan Meyer, Innsbruck, August 22, 1937, Archive Radiumforschung, Archive ÖAW.

[82] Marietta Blau, correspondence with Friedrich A. Paneth, History archive of the Max Planck Society, Berlin-Dahlem, Sec. III, Rep. 45 (Legacy Friedrich Adolf Paneth), File Nr. 17.

[83] Wolfgang Pauli (1900–1958), Viennese-born physicist, professor at the ETH Zurich and the Institute for Advanced Study, Princeton; Nobel Prize in Physics in 1945 for the discovery of the exclusion principle.

[84] Hans Bethe (1906–2005), German-born physicist, emigrated to England in 1933, since 1935 in the U.S.A.; 1943–1946, director of the theoretical physics division of the Los Alamos atomic bomb project.

[85] Erich Bagge (1912–1996), German physicist, work on atomic energy research during World War II at the Kaiser Wilhelm Institute, Berlin; after the war, University of Hamburg and Physikalisches Staatsinstitut Hamburg; later, University of Kiel, head of the Gesellschaft für Kernenergieverwertung in Schiffbau und Schifffahrt.

[86] Ellen Gleditsch, letter to Friedrich Paneth, Oslo, November 15, 1938, History archive of the Max Planck Society, Berlin-Dahlem, Sec. III, Rep. 45 (Legacy Friedrich Adolf Paneth), File Nr. 39.

[87] Albert Einstein, letter to William Edward Zeuch, Princeton, New Jersey, February 14, 1938, Albert Einstein Archive, Jerusalem, Mat. Nr. 52602.

[88] Rudolf Ladenburg (1882–1952), German physicist who had been on the board of directors of the Kaiser Wilhelm Institute for physics in Berlin. He emigrated to the U.S.A. in 1932 and was Albert Einstein's colleague and close friend at Princeton University.

[89] Albert Einstein, letter to Gustav Bucky, Princeton, New Jersey, February 14, 1938, Albert Einstein Archive, Jerusalem, Mat. Nr. 52602.1.

[90] Marietta Blau, letter to Elisabeth Rona and Berta Karlik, Copenhagen, March 15, 1938, Archive Radiumforschung, Archive ÖAW.

[91] Klaus Fischer, "Die Emigration deutschsprachiger Physiker 1933: Strukturen und Wirkungen," *Die Emigration der Wissenschaften nach 1933*, eds. Herbert A. Strauss et al. (Munich: Disziplingeschichtliche Studien, K.G. Saur, 1991), 25.

[92] Esperanza Verduzco Ríos, "La ESIME, Un refugio en México para Marietta Blau," *Investigación hoy* 95 (2000) 50.

[93] Franziska Loewe (Austrian Embassy, Mexico), letter to Robert Rosner, Mexico City, August 18, 2000.

[94] Marietta Blau, letter to Albert Einstein, Oslo, June 16, 1938, Albert Einstein Archive, Jerusalem, Mat. Nr. 52606.

[95] William Edward Zeuch, letter to Albert Einstein, Tlalpam, D.F., Mexico, February 9, 1938, Albert Einstein Archive, Jerusalem, Mat. Nr. 52601.

[96] Passports issued by the German government to Jewish people were stamped "J."

[97] Berta Karlik, letter to Ellen Gleditsch, Vienna, October 16, 1938 (copy), Sondersammlung Zentralbibl. Phys.

[98] Leopold Halpern was born February 17, 1925, in Vienna and earned his Ph.D. in physics at the University of Vienna in 1952. Between 1953 and 1974, he worked at the University of Vienna (assistant to Erwin Schrödinger), the University of North Carolina, the University of Stockholm, the University of Windsor (Canada), the University of Brussels, the University of Amsterdam, as well as Rensselaer Polytechnical Institute in Troy, New York, CERN in Geneva, the Niels Bohr Institute in Copenhagen, and the Institut Henri Poincaré in Paris. Since 1974, he was Senior Research Associate at Florida State University in Tallahassee, Florida (assistant to Maurice Dirac), with a leave spent at the Jet

Propulsion Laboratory in Pasadena, California. His main research was in the field of gravitational theory and its relation to elementary particle physics and quantum theory. He died on June 3, 2006, in Tallahassee.

[99] Stefan Meyer, letter to Fritz Paneth, Vienna, November 1, 1938, Archive Radiumforschung, Archive ÖAW.

[100] Esperanza Verduzco Ríos, personal communication to Robert Rosner, November 9, 2004.

[101] Marietta Blau, letter to Stefan Meyer, New York City, December 5, 1947, Archive Radiumforschung, Archive ÖAW.

[102] Richard Toeman, letter to Brigitte Strohmaier, London, November 19, 2000.

[103] Elizabeth Dresel, letter to Albert Einstein, Mexico D.F., September 11, 1939, Albert Einstein Archive, Jerusalem, Mat. Nr. 52609.

[104] Theo Schwarz, letter to Albert Einstein, Durango, Mexico, July 31, 1939, Albert Einstein Archive, Jerusalem, Mat. Nr. 52607.

[105] Marietta Blau, letter to Albert Einstein, Mexico D.F., June 16, 1941, Albert Einstein Archive, Jerusalem, Mat. Nr. 52635.

[106] Marietta Blau, letter to Stefan Meyer, New York City, December 9, 1946, Archive Radiumforschung, Archive ÖAW.

[107] Marietta Blau, letter to Eva Blau, Mexico City, July 19, 1941(?), courtesy of Eva Connors.

[108] Arnold Perlmutter, personal communication to Robert Rosner, June 2, 2000.

[109] Egon E. Kisch, *Läuse auf dem Markt* (Berlin: Aufbauverlag, 1985), 182.

[110] Christian Kloyber, "Einige Anmerkungen zum Exil österreichischer Intellektueller in Mexiko 1938 bis 1945," *Vertriebene Vernunft*, ed. F. Stadler (Vienna and Munich: Verlag für Jugend und Volk, 1987), 1010.

[111] Pierre Radvanyi, personal communication to Brigitte Strohmaier, January 21, 2000.

[112] The British landing in France mentioned in this letter took place on August 18, 1942, in Dieppe, and ended with a catastrophic defeat of the British: Of the six thousand soldiers who took part in the action, only two thousand returned.

[113] *Austria Libre*, March 1943.

[114] Marietta Blau reportedly said that her mother died from food poisoning after having eaten tortillas, Gerda Petkov, personal communication to Brigitte Strohmaier, February 20, 2001.

[115] Jules Romains (1885–1972), French writer, president of the International Pen Club 1936–1941.
[116] *Austria Libre*, May 1944.
[117] Karl Przibram, "Stefan Meyer (Obituary)," *Almanach Österr. Akad. Wiss.* 100 (1950) 340.
[118] Personnel file Dr. Georg Stetter, Archiv der Republik, Österreichisches Staatsarchiv, Vienna.
[119] Personnel file Dr. Gustav Ortner, Archiv der Republik, Österreichisches Staatsarchiv, Vienna.
[120] Josef Mattauch was born in 1895 in Moravian Ostrau, Silesia (then part of Austria). After his military service in World War I, he earned his Ph.D. under Felix Ehrenhaft at the University of Vienna in 1920. He spent two years with Robert Millikan at the California Institute of Technology (Pasadena), then was assistant and lecturer at the First Physics Institute in Vienna. With his student Richard Herzog, he developed the double-focusing mass spectrometer, a revolutionary instrument which opened a new era in the history of spectrometry of atomic masses and was essential for precision measurements. 1939 he was called to the Kaiser Wilhelm Institute (KWI) in Berlin-Dahlem by Otto Hahn, with whom he issued the *Isotope Report*, tables of the atomic species of the elements. In 1943 Mattauch became deputy director of the KWI but had to bear the full responsibility, as Hahn was interned in England. In 1947 Mattauch was appointed director of the institution, now the Max Planck Institute for chemistry. The institute was reestablished in Mainz and soon flourished as a Mecca for mass spectroscopy. Mattauch is also known for establishing a nuclear-physics mass scale with one-twelfth of the ^{12}C mass as a unit.
[121] Personnel file Dr. Hertha Wambacher, Archiv der Republik, Österreichisches Staatsarchiv, Vienna.
[122] Hertha Wambacher, "Kernzertrümmerung durch Höhenstrahlung in der photographischen Emulsion," *MIR* 435; *Sitzungsber. Akad. Wiss. Wien, Math. Naturwiss. Kl. IIa* 149 (1940) 157.
[123] Hertha Wambacher, "Über ein sicher identifiziertes Teilchen aus einer Höhenstrahlzertrümmerung," *Sitzungsber. Akad. Wiss. Wien, Math. Naturwiss. Kl. IIa* 154 (1945) 66.
[124] Georg Stetter and Hertha Wambacher, "Neuere Ergebnisse von Untersuchungen über die Mehrfachzertrümmerung von Atomkernen durch Höhenstrahlen," *Phys. Z.* 40 (1939) 702.

[125] Georg Stetter and Hertha Wambacher, "Versuche zur Absorption der Höhenstrahlung nach der photographischen Methode I: Zertrümmerungssterne unter Blei-Absorption," *Sitzungsber. Akad. Wiss. Wien, Math. Naturwiss. Kl. IIa* 152 (1944) 1.

[126] Hertha Wambacher and Anton Widhalm, "Über die kurzen Bahnspuren in photographischen Schichten," *Sitzungsber. Akad. Wiss. Wien, Math. Naturwiss. Kl. IIa* 152 (1944) 173.

[127] Hertha Wambacher, "Mehrfachzertrümmerung durch kosmische Strahlung; Ergebnisse aus 154 Zertrümmerungssternen in photographischen Platten," *Phys. Z.* 39 (1938) 883.

[128] Personnel file Hertha Wambacher, Archive of the University of Vienna, Vienna.

[129] Eva Connors, personal communication to Robert Rosner and Brigitte Strohmaier, July 25, 2001.

[130] Marietta Blau, letter to Stefan Meyer, Janesville, September 1, 1947, Archive Radiumforschung, Archive ÖAW.

[131] Marietta Blau, letter to Stefan Meyer, Janesville, August 4, 1947, Archive Radiumforschung, Archive ÖAW.

[132] Boris Pregel, letter to Marietta Blau, New York City, November 3, 1947 (copy), Sondersammlung Zentralbibl. Phys.

[133] Hendrik A. Lorentz (1853–1928), professor of physics in Leiden, Netherlands, principle of relativity, theory of electromagnetism; Nobel Prize in Physics, 1902.

[134] Max Planck (1858–1947), German physicist, professor at university of Berlin and director of the Kaiser-Wilhelm-Gesellschaft, Berlin; work on the nature and theory of heat, radiation law, Planck's constant; Nobel Prize in Physics, 1918.

[135] Enrico Fermi (1901–1954), Italian physicist, emigrated to the U.S.A. in 1938; 1944–1946, Manhattan Project, work on quantum mechanics, quantum electro dynamics, neutrons, nuclear reactions, β-decay; Nobel Prize in Physics, 1938.

[136] Isidor Isaac Rabi (1898–1988), physicist born in Rymanów (then Austria, now Poland); 1935, professor at Columbia University, New York City; 1940, deputy director, Radiation Lab, MIT; work on magnetism, molecular rays; Nobel Prize in Physics, 1944.

[137] Polykarp Kusch (1911–1993), German born physicist; 1949, professor at Columbia University, New York City; work on the structure of

atoms and molecules, atomic-ray resonance method; Nobel Prize in Physics, 1955, together with W. Lamb.

[138] Willis E. Lamb (born 1913, California), professor of physics at Columbia University, Stanford University, Harvard University, University of Oxford, England, Yale University; work on maser, laser, β-decay, neutron scattering, molecular-ray resonance method; Nobel Prize in Physics 1955, together with P. Kusch.

[139] Marietta Blau, letter to Stefan Meyer, New York, March 14, 1948, Archive Radiumforschung, Archive ÖAW.

[140] Marietta Blau, letter to Hans Thirring, New York, July 26, 1948, Archive Radiumforschung, Archive ÖAW.

[141] Peter Havas, born in Budapest in 1916, studied theoretical physics at the Technical University in Vienna and from 1938 on at Columbia University. From 1946 to 1965, he worked at Lehigh University, Bethlehem, Pennsylvania. He died near Philadelphia in 2004.

[142] Web Site of The Nobel Foundation, http://www.nobel.se/laureate.

[143] Erwin Schrödinger, letter to the Nobel Committee, Dublin, January 8, 1950, Nobel Archive at the Center for History of Science, Royal Swedish Academy of Sciences.

[144] C.F. Powell and G.P.S Occhialini, *Nuclear Physics in Photographs* (Oxford: Oxford University Press, 1947).

[145] Arnold Perlmutter, letter to *Physics Today*, August 1998, 81.

[146] Walter Thirring was born in 1927 in Vienna, son of Hans Thirring; he earned his Ph.D. in physics at the University of Vienna in 1949. After nine years as visiting professor at the Dublin Institute for Advanced Studies, the Max Planck Institute for physics in Göttingen, the ETH Zurich, the University of Bern, the MIT, Cambridge, the University of Washington, Seattle, and the Princeton Institute for Advanced Studies he was appointed professor at the University of Bern in 1958. He became full professor and head of the Institute for Theoretical Physics at the University of Vienna in 1959. From 1968 to 1971, he was director of the theoretical department of CERN. His main fields of research were quantum electrodynamics and quantum field theory. He has been professor emeritus since 1995.

[147] Walter Thirring, letter to Wolfgang Kerber, October 7, 2003.

[148] Protokoll vid Kungl. Vetenskapsakademiens sammankomster för behandling av ärenden rörande Nobelstiftelsen, år 1950, Verdict from

July 1, 1950, Nobel Archive at the Center for History of Science, Royal Swedish Academy of Sciences.

[149] Axel E. Lindh (1888–1960), 1932–1937, professor at the Chalmers Tekniska Institutet in Göteborg; 1937–1954, professor and director of the Physics Institute at the University of Uppsala; 1935–1954, member of the Nobel Prize Committee for Physics; 1941–1945, head of the Swedish National Committee for Physics, Militärfysiska Institutet; 1942–1947, president of the Swedish Physical Society.

[150] Marietta Blau, letter to Eva Blau, Bath, January 1953, courtesy of Eva Connors.

[151] Margarethe Heinrich, telephone communication to Brigitte Strohmaier, March 29, 2000.

[152] Anny Schlemko-Frantz, telephone communication to Brigitte Strohmaier, February 26, 1998.

[153] Georg Stetter in: "Personal- und Hochschulnachrichten," *Österr. Chemiker-Zeitung* 51 (1950) 234.

[154] Georg Stetter and Hans Thirring, "Hertha Wambacher ✝," *Acta Phys. Austriaca* 4 (1951) 318.

[155] Georg Wagner, "In Memoriam Hertha Wambacher," *Natur und Technik* 4 (1950) 142.

[156] Georg Wagner (1919–1977), Austrian chemist; in 1950, he held the title of an associate professor for analytical chemistry at the Technical University in Vienna.

[157] Gustav Ortner, "Dr. H. Wambacher (Obituary)," *Nature* 166 (1950) 135.

[158] Berta Karlik, "1938–1950," *Sitzungsber. Österr. Akad. Wiss., Math. Naturwiss. Kl. IIa* 159 (1950) 35.

[159] Stefan Meyer, letter to Marietta Blau, Bad Ischl, July 27, 1947, Archive Radiumforschung, Archive ÖAW.

[160] Eduard Haschek, born 1875, Vienna, died 1947, Klosterneuburg, studied physics at the University of Vienna starting in 1893; dissertation with F. Exner; 1897, Ph.D.; 1919, associate professor; 1931, full professor of physics at the University of Vienna; 1934, Lieben Prize for research in the field of chromatics.

[161] Gilberto Bernardini (1906–1995), Italian physicist, worked on cosmic rays, president of the Italian National Nuclear Physics Institute (INFN), later proton synchrotron research director at CERN.

[162] Marietta Blau, letter to Berta Karlik, Patchogue, New York, March 4, 1951, Archive Radiumforschung, Archive ÖAW.
[163] Sol Pearlstein, communication to Brigitte Strohmaier and Alfred Chalupka, May 12, 1980.
[164] Eva Connors, letter to Brigitte Strohmaier, September 19, 2000.
[165] Hannelore Sexl, communication to Brigitte Strohmaier, March 27, 2000.
[166] Marietta Blau, letter to Berta Karlik, Bristol, January 13, 1953, Archive Radiumforschung, Archive ÖAW.
[167] Hans Thirring, letter to the Nobel Committee for Physics in Stockholm, January 29, 1955, Sondersammlung Zentralbibl. Phys.
[168] Erwin Schrödinger, letter to Max Born, October 24, 1960, in W. Moore, *Schrödinger – Life and Thought* (Cambridge, New York: Cambridge University Press, 1989), 479.
[169] Marietta Blau, letter to Berta Karlik, Wembley, Middlesex, September 26, 1955 (copy), Sondersammlung Zentralbibl. Phys.
[170] Marietta Blau, letter to Berta Karlik, Miami, March 24, 1956, Archive Radiumforschung, Archive ÖAW.
[171] Marietta Blau, letter to Berta Karlik, Coral Gables, probably fall 1956 (copy), Sondersammlung Zentralbibl. Phys.
[172] Otto Blau, letter to Leopold Halpern, Gentilino near Lugano, January 22, 1977 (copy), Sondersammlung Zentralbibl. Phys.
[173] Marietta Blau, letter to Berta Karlik, Coral Gables, September 16, 1957 (copy), Sondersammlung Zentralbibl. Phys.
[174] Collection of Academy Direction Nr. 363, Working Group History of the Berlin Academy, Berlin-Brandenburg Academy of Sciences, Berlin.
[175] Marietta Blau, letter to Berta Karlik, Coral Gables, probably at the end of 1959 (copy), Sondersammlung Zentralbibl. Phys.
[176] Blau obtained this apartment (in Charasgasse 8 in the third district) with the help of Maria Jacobi, member of the Vienna city council; the house in which the apartment was located belonged to the Wiener Verein, a funeral and insurance company.
[177] Hannelore Sexl, communication to Reinhard Schlögl, February 24, 1999.
[178] Marietta Blau, letter to Arnold Perlmutter, Vienna, August 8, 1960, courtesy of Arnold Perlmutter.

[179] K. Przibram, L. Flamm, F. Machatschki, H. Nowotny, F. Regler, E. Schmid, and F. Wessely, letter to the Section for Mathematics and Science of the Austrian Academy of Sciences, Vienna, March 18, 1961, Archive ÖAW.

[180] Protocol of the election session of the Austrian Academy of Sciences, May 30, 1961, Archive ÖAW.

[181] Richard Meister and Fritz Knoll, letter to the Federal Ministry of Education, Vienna, April 28, 1961, Archive ÖAW.

[182] *Almanach Österr. Akad. Wiss.* 112 (1962) 242.

[183] Hugo Zwins, Gerhard Czapek, Helmut Romatnik and Gottfried Kellner.

[184] Gerda Haider, *Untersuchung der Wechselwirkungen von 10 GeV/c μ^--Mesonen in Emulsionen*, Dissertation, University of Vienna (1964).

[185] Marietta Blau, letter to Arnold Perlmutter, Vienna, November 1, 1963, courtesy of Arnold Perlmutter.

[186] Peter Hille, communication to Brigitte Strohmaier, November 13, 1999.

[187] Hannelore Sexl, telephone communication to Brigitte Strohmaier, January 26, 2000.

[188] Wolfgang Breunlich, communication to Brigitte Strohmaier, February 17, 2000.

[189] Marietta Blau, letter to Leopold Halpern, Vienna, April 28, 1964 (copy), Sondersammlung Zentralbibl. Phys.

[190] Waltraut Vonach, née Patzak, communication to Brigitte Strohmaier, November 9, 1999.

[191] Heinz Felber, communication to Brigitte Strohmaier, December 13, 1999.

[192] Anna Saulich, telephone communication to Brigitte Strohmaier, December 22, 1999.

[193] Gerda Petkov, communication to Reinhard Schlögl, February 10, 1999.

[194] Marietta Blau, letter to Arnold Perlmutter, Vienna, November 10, 1964, courtesy of Arnold Perlmutter.

[195] Pschyrembel, *Klinisches Wörterbuch*, 256th Edition (Berlin, New York: Walter de Gruyter, 1989), 80.

[196] Ortrud Grön, *Das offene Geheimnis der Träume. Ein Leitfaden* (Freiburg: KORE-Verlag, 1998).

[197] Hospital report on Marietta Blau (1966–1970), Lainz Hospital, Vienna.
[198] *Amtsblatt der Stadt Wien*, Nr. 40, May 20, 1967, p. 14.
[199] Berta Karlik, letter to the professors' board of the philosophical faculty of the University of Vienna, December 2, 1968, Archive of the University of Vienna, Vienna.
[200] Rudolf Hanslik, letter to Marietta Blau, March 11, 1968, Archive of the University of Vienna, Vienna.
[201] Marietta Blau, letter to Rudolf Hanslik, March 16, 1968, Archive of the University of Vienna, Vienna.
[202] Marietta Blau, letter to Rudolf Hanslik, July 16, 1968, Archive of the University of Vienna, Vienna.
[203] Announcement of Marietta Blau's death (copy), Sondersammlung Zentralbibl. Phys.
[204] Berta Karlik, "Bericht aus dem Institut für Radiumforschung und Kernphysik," *Almanach Österr. Akad. Wiss.* 120 (1970) 200.
[205] Leopold von Pfaundler (1839–1920), Austrian physicist, professor at the University of Graz.
[206] Karl Przibram, *Verfärbung und Lumineszenz* (Vienna: Springer, 1953).
[207] Berta Karlik and Erich Schmid, *Franz S. Exner und sein Kreis* (Vienna: Austrian Academy of Sciences, 1982).
[208] C.-S. Wu and S.A. Moszkowski, *Beta Decay* (New York: Interscience Publishers, 1966).

Marietta Blau – Teacher and Friend

Contacts with Marietta Blau in Vienna, 1935–1937 and after 1960
Hanne Ellis-Lauda, Vienna[1]

I met Marietta Blau when I was looking for a place to complete the thesis for my doctoral degree. One of my friends knew that Marietta Blau was looking for students who would work for her. The main part of her work at that time was the study of traces which certain atomic particles produced in special photographic emulsions. After a brief interview in autumn 1934, I started working with her. She introduced me to the photographic method and later on asked me to study the fading of the latent image in photographic emulsions stored at various conditions, e.g., in a vacuum, at different temperatures, etc. The time I needed for this work turned out to be longer than expected because the vacuum did not hold through the entire summer and a series of experiments had to be repeated. While Marietta Blau allowed me to work independently, she was always very kind and helpful when asked for advice.

In 1938 Marietta Blau could not continue her scientific work in Austria. Her first student in photographic emulsion work, Hertha Wambacher, who had continued working with her as a colleague after completing her Ph.D. in 1932, had, however, for some time been a member of the National Socialist Party and was highly anti-Semitic. She displayed a great dislike for Blau and her unusual traits of modesty combined with self-confidence.

Blau, on the other hand, was on good terms with Elisabeth Rona and friends with Berta Karlik, although she did not have much to do with her as far as work was concerned.

Marietta Blau left Austria for Oslo, Great Britain, and finally in 1939 for Mexico City, but the job offered to her gave her no chance for scientific work. In 1948 she took a job at Columbia University in New York, where she was able to resume her work with photographic emulsions.

I received my doctorate in 1937 and started working in the electrotechnical industry in Vienna and Berlin.

Hertha Wambacher was barred from the university for political reasons after the war, and Berta Karlik, then head of the Radium Institute, asked me to continue work with photographic emulsions from 1947 onwards.

In 1951 I was offered a Canada-UNESCO-fellowship which allowed me to visit all the universities in Canada where emulsion work was being done. I contacted Marietta Blau, and she invited me to come to New York where an international meeting of the American Physical Society was being held and I had the unique opportunity to meet all the famous physicists of the time.

When Marietta Blau moved back to Austria in 1960, she was given the chance to work again at the Radium Institute in her main field of interest, the photographic emulsion method.

She seemed lonely in her private life. Only Berta Karlik and I were left from among her friends before the war, and Prof. Karlik was very busy as the head of the Radium Institute.

Blau liked to have dinner with my young son (then four years old) and me. She brought him books, which he loved and still reads to his own children.

Since she could not drive any more because of her eyesight and had no car, we often went in my car to the Vienna Woods. Heiligenkreuz with its large forests was her favorite place for tea.

After she unfortunately developed cancer, I went to see her in her apartment and finally in the hospital, where she patiently endured her suffering without complaint. I saw her for the last time a few days before she died in 1970.

MARIETTA BLAU – MY TEACHER IN MEXICO IN 1943
Pierre Radvanyi, former Research Director of CNRS, Orsay, France[2]

It was towards the end of the year 1942 in Mexico. The school year begins in February there and ends at the beginning of November. I was to attend the final grade of the French lycée of Mexico in the section for elementary mathematics in which I was the only student enrolled. I had the required books at my disposal, but there was no regular professor at the lycée at that time who could have given me instruction in physics and chemistry. Among the emigrants like us who had been forced to leave Europe, my mother discovered Marietta Blau, who had come from the Radium Institute in Vienna. She asked her whether she was willing to tutor me in these subjects. Marietta Blau accepted, and that is how I made her acquaintance.

During 1943, I went to see her regularly at her apartment. Our conversations were either in German or French. She was of small stature; her black hair was in a bun, and her luminous black eyes sparkled with intelligence. Usually she wore dark clothes. At that time, she lived on one of the upper floors of a large building. The first time I had visited her she was living in another apartment with her equally diminutive mother, who was amiable and reserved.

The second time I came to see her, Blau showed me an electroscope and explained how it functioned by using a piece of naturally radioactive rock. Once she said to me: "Some chemical elements are like living beings: They arise and disappear." This idea made a deep impression upon me and has remained engraved in my memory.

During our meetings, I would ask her to explain questions I had not understood, and she quizzed me on the chapters of the books I had studied. In this way, we covered mechanics, optics, electricity, magnetism, wave phenomena, and also some principles of organic chemistry. At the end of that year, I received my baccalaureate with the grade "very good."

Marietta Blau was lonely in Mexico. She used to talk with great affection of the Radium Institutes in Vienna and Paris. It was virtually impossible for her to find employment in Mexico suitable for carrying out scientific investigations. Finally, a connection to Sandoval Vallarta's group evolved, but this was not a real job.

I saw her once more the following year in a laboratory belonging to M. Sandoval Vallarta's cosmic-ray group, when she worked at the national Mexican polytechnical institute. With the help of one or two students, she repaired Geiger counters.

After the war ended, I went to France again to continue my studies on a scholarship in Paris.

COLLABORATION WITH MARIETTA BLAU AT COLUMBIA
UNIVERSITY IN 1949
Martin M. Block, Northwestern University, Evanston, Illinois[3]

Indeed, I did work with Marietta Blau at Columbia in the late 1940s. I was a graduate student and she was a research associate. As a graduate student, I was in charge of the nuclear emulsions section. Of course, none of us knew anything about emulsions – it took Blau to teach us the techniques of developing, scanning, etc. In this she was indispensable ... We were still building the Nevis Cyclotron (in those days, grad students really built the equipment, tested it, etc., since we had no engineers or post-doctoral students) and for practice exposed some nuclear emulsions in balloons. We came across a strange event, which we published in 1949. It took until 1999 for me to work again on cosmic rays – I published a paper in *Physical Review Letters* on December 13 that year. I can't tell you anything about her work at Brookhaven. I do know that she was treated rather shabbily at Columbia – but who wasn't during that time.

MY INTERACTION WITH MARIETTA BLAU AT COLUMBIA
UNIVERSITY IN 1950
Seymour J. Lindenbaum, Brookhaven National Laboratory, Upton, New York[4]

After the war, I simultaneously was employed as a full-time staff member of the Nevis Cyclotron Laboratories of Columbia Universities (1946–1951) and was pursuing my Ph.D. in physics at Columbia.

I was initially assisting Professor Rainwater (in effect, a deputy director) under Professor Booth (the director) in developing the Radio Frequency System for the approximately 400-MeV Nevis Cyclotron design and construction project. However, my duties were soon broadened in this small scientific staff to include most aspects of the Cyclotron Project, including planning elements of the future research program.

Marietta Blau was a senior staff member brought in to prepare an emulsion program for the cyclotron because she was the well-known outstanding pioneer in the emulsion technique and its application for studying nuclear stars.

I was asked to look into her program and help where I could. I found Marietta to be almost unbelievably well versed in every aspect of the emulsion technique, even though my expectations were high because of the outstanding reputation that she had.

She set up a very complete emulsion laboratory with scanners whom she trained herself. Marietta also taught me the emulsion technique at a sophisticated level, which was impossible to attain through the literature I read.

It was obvious to her that the high data rates an FM [frequency modulated] cyclotron would generate would overwhelm human scanners. She felt that as much automation as possible should be developed for the scanning and analysis programs because doing so would eliminate most of the biases introduced by human scanning. I totally shared her conclusion, and since I assumed that the first Nevis Cyclotron experiment would be completed most quickly with emulsions, I chose to study one of these central topics of the day for my thesis, namely the process or processes that produce cosmic-ray stars.

Thus I began to collaborate extensively with Marietta and an engineer, R. Rudin, on the development of a semi-automatic device for analyzing events in nuclear emulsions. This program was successfully completed and the results published.[B62, B65]

I believe if emphasis had not quickly shifted to the bubble chamber, and soon thereafter to the electronic bubble chambers, which I later pioneered, that this semi-automatic device would have become quite important. In any event, it was the first idea which was successfully developed to automate visually observed events and likely influenced future developments.

Marietta Blau, in addition to being an outstanding scientist, was a warm and caring individual whose company was always very much enjoyed by me and her colleagues, and although a quiet individual, she certainly stimulated me and other colleagues greatly.

I went on to complete my thesis – the first Nevis experiment – by exposing emulsions sensitive to minimum-ionization tracks to an internal 350–400 MeV proton beam and demonstrated that a cascade of individual nucleons was actually the major element in nuclear-star production. This was in contrast to the most popular belief at the time, that the Fermi statistical theory explained the major mechanism of star formation. The only role of the Fermi statistical theory was to cause a thermodynamic evaporation due to rearrangement of the holes in the Fermi sea produced by the nucleonic cascade.[5]

Although Marietta Blau did not participate in this extensive, important program, it was the laboratory she developed which was used for it.

MEMORIES OF A LENGTHY COLLABORATION BEGINNING IN 1956
Arnold Perlmutter, University of Miami, Coral Gables, Florida[6]

To the best of my recollection, Marietta Blau came to Coral Gables as an associate professor in autumn 1956. I had arrived in February of that year as an assistant professor and continued my interest in solid-state physics by investigating the optical properties of the semiconductor GaAs.

Because I do not have a very good memory for detail, I cannot recount the specifics of my meeting with Blau, but I do recall that we had a nearly instant rapport and that I responded without hesitation to her invitation to collaborate with her on photographic-emulsion research in high-energy physics. She also recruited three other colleagues, including Claude F. Carter. The other two did not remain with the group.

It was not simply on a whim that I answered her invitation because, in the previous few years before and after completing my doctorate in 1955 in solid-state physics, I had made a serious attempt to study theoretical physics, particularly nuclear physics, independently with some guidance from a former teacher at New York University. Before I made any substantial progress on my own, the thought of having the opportunity to do high-level experimental physics with such a distinguished mentor was exciting.

Marietta Blau was a rather small person, perhaps 5 feet 2 inches tall (158 cm) and quite slender, with a sweet, kindly expression. Her head was barely visible over the steering wheel of her little Plymouth coupe, and although she was not a very skilled driver, she negotiated the trip from New York to Miami several times before the days of interstate highways. The initial impression she made was that of a fragile person who could be blown over by a breeze. I would say that she was quite good looking but presented herself in a very modest, self-effacing manner. She spoke deliberately, slowly, and softly, and her English, though slightly accented, was polished. She was well versed in the classics, literature, and the arts. We attended many musical events together, especially performances of chamber-music ensembles on tour. I recall that we once attended a presentation of Verdi's *Requiem*, by the Miami Symphony (then the University of Miami Symphony) and that we were so mesmerized that we sent a warm letter of appreciation together to the conductor, John Bitter (a brother of the well-known authority on high-field magnets, Francis Bitter). It was the first time I had heard the *Requiem*.

When she arrived in Miami, she found several former friends and students from Vienna, namely Fritz Koczy and Elisabeth Rona, both of them at the Institute of Marine Sciences at the University of Miami. Koczy later became director of the Marine Institute, and he and I became close friends until his untimely death in the 1970s, I believe.

My oldest son, Bernard, was three years old when we arrived in Miami, and Joseph was born in 1957, just after Marietta arrived. She became a close family member, showering the children and my first wife, Ruth, with gifts. My older son reminds me that she gave them an Indian tent to sleep out in when he was about six years old. We often entertained each other in our homes. She met my parents on their visits to Miami, and they all enjoyed each other's company greatly. She grieved compassionately when my mother died in 1960, shortly after Marietta's return to Vienna.

Initially, we were housed in cramped quarters in the Physics Department, at that time in wooden shacks which were used during World War II for the training of Air Force pilots. The University of Miami then rented space for us on the ground floor of an old apartment building in Coral Gables, about two miles away from the Main Campus. We had about eight or ten rooms, mostly devoted to scanning microscopes, an equipment room, and of course the ubiquitous coffee lounge. One large room was reserved for a multiple scattering microscope, with a large base and a massive stage. It was the massive base of concrete blocks for this microscope that required that our laboratory be located on the ground floor of the building. An interesting dividend of our needs (the microscope and the fragile emulsions) was that the entire facility had to be air conditioned (at that time by window units). This was an unusual luxury in those years in Miami, which suffers sweltering temperatures for half of the year.

Marietta Blau was a most effective teacher, giving courses in electromagnetism and nuclear physics, among other subjects, to advanced undergraduates and graduate students. At that time we did not have a Ph.D. program but offered a Master of Science degree. I believe that because of her slight stature and her gender, she was not afforded the respect to which she was entitled. She fought with the administration and the chairman of the Physics Department about the use of overhead from her federal grants. After one complaint to another colleague, the theorist Behram Kursunoglu, I recall that the latter called the chairman an idiot. She replied in her soft, plaintive voice, "Is that all he is?"

I really cannot recall how I learned how to work with emulsions, but I do know that Marietta Blau was a wonderful teacher in the laboratory. We recruited housewives and students, in the tradition of Cecil Powell, to be scanners for the emulsion pellicles and, in the cases of the more gifted assistants, allowed them to take precision measurements. She of course instructed me in the theory of ionization measurements, multiple scattering, and range-energy relations. All of the faculty scanned, supervised the scanners, and made the precise measurements as needed.

As the bubble chamber, spark chamber, streamer chamber, and digitized chambers evolved in the 1960s and 1970s, emulsion research became less important for the accumulation of statistics. One of the reasons is that the bubble chambers and spark chambers could be composed of elemental substances, such as hydrogen, deuterium, helium, or other interesting substances, and that the measurements could be automated to the extent that made the accumulation of data far more rapid and efficient. In modern applications, electronics, solid-state detectors, and other devices further increased the effectiveness of detectors. However, emulsions remain an interesting detector in special situations, especially where spatial resolution and continual exposure are needed.

In April 1960, Marietta left by train from Miami for New York, and then traveled on to Vienna. My family and the Carters bade her an emotional farewell.

I return now to the subject that must have been the source of great pain and frustration in Marietta Blau's professional life, namely the official neglect of her role in the discovery of the pion. She was too proud and private to reveal this disappointment to me openly, but I do recall that she had great disdain for C.F. Powell, who was awarded the Nobel Prize in Physics 1950 "for his development of the photographic method of studying nuclear processes and his discoveries regarding mesons made with this method." Incorrectly, she attributed the discovery of the π^--meson to Perkins and that of the π^+-meson to Lattes, Occhialini, and Powell, which is corroborated by reading her *Acta Physica Austriaca* article[B64] and her article in Yuan and Wu's book.[B79] I now remember that of Powell's group she thought most highly of Lattes, who co-authored the decisive publication with Powell. She considered Lattes an infant prodigy.

After Marietta had returned to Vienna, I carried on our common research projects. At first these were our results of interaction of K^--mesons with emulsions. Then I exposed emulsion plates to the 800 MeV/c K^--meson beam at the Bevatron in Berkeley; I evaluated them in Miami. In Trieste (1961/62), I collaborated with the bubble-chamber group, which was scanning photographs of the antiproton beam at CERN. With the help of three scanners, I found two heavy hyperfragment decays produced by the fast K^--beam. I also performed phenomenological computations on antiproton-proton interactions, and found decent agreement with the experimental results then available.

In the spring of 1962, my family and I traveled from Trieste to Vienna, where we visited with Marietta Blau at her apartment. She was, of course, a gracious hostess, but the main purpose of the visit was to finish our hyperon paper.[B80] She expressed great displeasure at her treatment by the faculty at the Institut für Radiumforschung and her inability to be involved in productive work. She was clearly not in the best of physical health.

I extended my leave from the University of Miami by going to the Weizmann Institute of Science in Rehovoth, Israel. There I collaborated (1962/63) with a group that had been studying the interactions of K^--mesons with emulsion nuclei with the production of ($\Sigma^\pm + \pi^\mp$). We combined their results with ours from Miami and carried out a detailed analysis of the results. We found that the Coulomb effect on the energies of the charged Σ-hyperons and charged π-mesons was observable and had an effect on the $\Lambda(1405)$ resonance. In 1963 I returned to the University of Miami, where I worked on a phenomenological treatment of particle scattering. Later I took part in experiments using spark chambers designed to measure the magnetic moment of the Σ^+-hyperon at Argonne National Laboratory and in experiments on high energy polarized proton-proton scattering at Argonne National Laboratory and Brookhaven National Laboratory, where the particle detectors were scintillation counters.

I mention these endeavors mainly to emphasize my indebtedness to the influence and legacy of Marietta Blau over the rest of my life. She introduced me to particle physics, and even if I am dissatisfied with the qualitiy of my achievements, I am grateful for being a participant in interesting experiments and in having established a worldwide circle of friends and associates in this exciting field.

RECOLLECTIONS OF MARIETTA BLAU IN MIAMI, 1960
Sylvan C. Bloch, Professor Emeritus, University of South Florida, Tampa[7]

I was privileged to be one of Professor Blau's graduate students – perhaps her last one. Under her direction, I earned a master's degree while participating in her research project at the University of Miami funded by the Air Force Office of Scientific Research. The research was based on the use of nuclear emulsions to study fundamental particle interactions. This work resulted in one of my first publications, which happened to be one of Professor Blau's last publications: "Studies of Ionization Parameters in Nuclear Emulsions."[B76]

I was also a student in Professor Blau's course in nuclear physics. She was an excellent teacher, very demanding and respected by her students. As in her research, she had high standards and expected the best from her students.

Professor Blau was no less demanding of her colleagues. I was in her office one day when she was berating another professor, saying, "You know nothing about fundamental particles!"

ACQUAINTANCE WITH MARIETTA BLAU IN THE UNITED STATES AND IN VIENNA
Leopold Halpern, Florida State University, Tallahassee[8]

Blau was a friend of Herta Leng's, the professor with whom I worked during my first Fulbright fellowship. Leng, like Marietta Blau, was an Austrian physicist who had worked as a volunteer at the Vienna Radium Institute (while earning her living as a grammar-school teacher). She emigrated and became a professor at the Rensselaer Polytechnical Institute in Troy, New York. She created an excellent laboratory for students while continuing her outstanding research. It was there that I first met Marietta Blau in 1952. Leng had invited her to Troy to give a lecture, and afterwards we spoke. At the time, Blau was working at the Brookhaven National Laboratory. As soon as she noticed my interest in Brookhaven, to which foreigners had limited access due to restricted visitors' permits, she invited me to visit there. Arranging the visit was complicated since everything at Brookhaven was either confidential or classified because it involved nuclear research. Blau was very helpful.

With the photomultiplier, Marietta Blau developed the first scanning machines for automatic determination of particle tracks. Her work at Columbia University and in Brookhaven was extremely fruitful and valuable, but she felt quite isolated there and was concerned about her eyesight, which after years of handling radioactive material was threatened by cataracts. I got to know her better and took a sympathetic interest in her plight whenever I met her. Later she felt unable to continue her normal work without undergoing a cataract operation. But she could not afford this surgery in the United States, prompting her to return to Austria. Her relatively short, albeit exceedingly successful, period of work in the United States had earned her a pension of only two hundred dollars per month by the time she returned to Vienna. Medical expenses in Austria were relatively low, so she decided to have the surgery done there.

From then on I saw her every year. I would not have become so interested in her as far as the history of science is concerned had I not noticed how much information about her had been falsified or suppressed.

I met her in Vienna in 1962, and in the following years when I came on visits to Vienna, I went to see her in her apartment. We often talked for hours. She told me a lot about her experiences and her career.

She was permitted to earn hundred dollars monthly tax free in addition to her two hundred dollars pension and intended to do so, without quite knowing how. At that time, I was staying in Copenhagen and came to Vienna occasionally. When I visited her, she mentioned her problem. She had considered trying to earn that amount of money either as an editor of a scientific paper or by co-authoring CERN publications. With the help of Mr. Rozental, the administrative director in Copenhagen, and that of the Austrian physicist at CERN, Viktor Weisskopf, I looked into these possibilities. Incredibly, her wish was turned down by CERN, even though the most important people supported her, among them Otto Frisch, Lise Meitner's nephew. He, too, had no success; all he was able to arrange was a one-time scholarship.

MARIETTA BLAU AND THE VIENNA PLATE GROUP
Brigitte Buschbeck, Institute for High Energy Physics, Austrian Academy of Sciences, Vienna[9]

After I finished my studies in 1959, Professor Walter Thirring, who had just accepted an appointment in Vienna, asked me whether I would take charge of a small high-energy physics group there. At that time, I knew nothing about high-energy physics; therefore, Professor Thirring sent me to Bern and then to CERN in Geneva for a year. When I came back, I took over the group consisting of four students and four female scanners who evaluated the photographic plates under the microscope and also manually evaluated bubble-chamber photographs produced at CERN.

Since I did not really have the competence to plan four dissertations for young students even after a year at CERN, Professor Thirring took the opportunity to ask Professor Blau, who had just returned after her retirement, to assist us. I remember very clearly how she entered. She was a short, sweet, modest, older lady, rather shy. When she saw us and the way we were working, she wrung her hands in dismay, saying it was impossible to work under such conditions, as we were inadequately equipped. We had been so proud of ourselves, though, because we were just launching our project. To me, her statement that we would be incapable of producing a dissertation was like being thrust under a cold shower. At the time, I responded: "What can I do? I have been placed here and I have to do what I can." She reacted very kindly, but for me at that time she was unapproachable and I took the criticism personally, "How can you work here? You lack apparatus." After all, she came from the United States. I was very young then and felt uneasy due to her critical attitude towards our group.

Later, Professor Blau supervised a fifth doctoral student, Gerda Haider (later Petkov) and, together with her, produced a wonderful thesis using a special microscope we bought. It may not have been the best, but it was a good beginning. Professor Blau was given a small room at the Radium Institute where she and Gerda Haider worked together. She was thus separated from us completely and dedicated all her time to this one dissertation. In the end, Dr. Haider gave a very interesting talk under the supervision of Professor Blau.

IN SCIENTIFIC DIALOGUE WITH MARIETTA BLAU
Herbert Pietschmann, Institute for Theoretical Physics,
University of Vienna[10]

At the beginning of the 1960s, just after I returned from CERN, I was assigned to the plate group as house theoretician (see contribution by B. Buschbeck). I was to assist those colleagues evaluating track photographs by the application and interpretation of theoretical principles.

One day, somebody – I do not recall who – asked me whether I would be willing to help an older female colleague from the Radium Institute. Naturally, I agreed, but did not know that it was Marietta Blau whom I was going to assist. I admit that at that time, as a very young theoretician, I had no idea of Marietta Blau's standing in nuclear and particle physics. As a result, our meeting was completely free and uninhibited.

Marietta Blau came to my desk with someone else (I do not remember with whom). I was surprised that so fragile and small a person with such a soft, shy voice could exude such authority. Her movements were cautious and slow, as if to prevent anything from escaping her attention. She sat down at my desk and started asking questions but seemed to have trouble finding the right words. I recognized immediately that she was using English for all the physics terms, which at that time was less common than nowadays. Some humorous combinations of language resulted, like "x divided by zwei" and things like that.

Her first few questions were so basic that I doubted her expertise, but she soon began asking more profound questions I had difficulty answering. Soon after that, I had a similar experience when I was invited by Werner Heisenberg in 1963 to give a talk on my work in current algebra at his institute. Heisenberg's first question was that of a beginner, but he soon afterwards directed his questions directly at the weak points in my derivation and displayed a speed of comprehension that I had never experienced before.

Evidently, this was the way in which Marietta Blau (and Werner Heisenberg) approached a problem; in order to understand something entirely, even the most basic detail needs to be clarified.

Another similarity comes to my mind when I think of Marietta Blau. About the same time, Lise Meitner visited our institute. (On this occasion, we even bought a new tea set because the old one wasn't suitable for guests!) Lise Meitner, likewise an older woman, was not as fragile and shy as Marietta Blau, but there were parallels. In retrospect, these similarities appear to me to express a fate that the two "grandes dames" of Austrian physics shared. Both were passionately devoted to their discipline, and both were denied the highest acknowledgment of their work.

JOINT LABORATORY WORK AND A TRIP TO SWITZERLAND WITH MARIETTA BLAU IN 1961
Hannelore Sexl, Commission for the History of Science, Mathematics and Medicine of the Austrian Academy of Sciences, Vienna[11]

In 1960 I applied for a dissertation at the Radium Institute. Berta Karlik proposed that Gerda Haider and I work together with Marietta on the photographic method in combination with the evaluation of bubble-chamber exposures. I immediately had complete confidence in Marietta Blau.

Blau came to the institute every day and attentively looked after us. I guess she was somewhat lonesome and quite happy to have something to occupy her. We learned how to evaluate bubble-chamber exposures from her and also from visitors who were in close contact with Marietta, such as Professor Winzeler from Bern and Professor Morrison from CERN.

Thanks to these contacts, we undertook a trip to CERN in Geneva. We spent nine days with Blau and saw many sights together, such as Mont Blanc. We were on personal terms with her. When we asked her questions about her private life, even about her admirers in her youth, she always responded but never in great detail. Also, she never talked about why she went to Göttingen. She did tell us, though, that things in Göttingen were not quite the way they used to be in Vienna. Stefan Meyer had been so kind and by no means as patriarchal as Pohl in Göttingen, where a real hierarchy existed. Pohl was at the top like a god. The situation at the Vienna Radium Institute was completely different. According to her, Stefan Meyer was charming in his way. She had a high opinion of him and often spoke about him to us.

She never spoke about her emigration but often did about her teaching in Mexico and Miami. She never mentioned how she came to emigrate or how she felt about it. Maybe we just didn't ask directly enough about it.

It was a wonderful journey filled with a positive feeling of collaboration and confidence, which I rarely felt again in my life. For instance, we stopped in Innsbruck, where Marietta Blau wanted to visit the imperial church to see the *Mander*, because she had not been there in years. But she repeatedly emphasized, "If you don't want to, we don't need to go there." [The *Mander* are twenty-eight statues in bronze, more than life-size designed by Albrecht Dürer in the sixteenth century. They are located in the church of the former Habsburg court and represent historical personalities.]

Following an invitation by Professor Houtermans, we went to Bern. Marietta Blau gave two talks there, one on the Blau-Wambacher stars and one on hyperfragments. She also gave several talks in Geneva. At CERN, we took part in a long guided tour. I had the feeling that Marietta enjoyed being together with other physicists. The last evening of that trip we spent at Mondsee, where my grandfather owned a small villa which she liked. Afterwards, she sent my grandfather a book as a thank-you. My grandfather was touched by the generous gift. She was always extremely generous.

At the Radium Institute, she had a small room, in fact the same room which Berta Karlik used as a professor emerita. In the 1960s, Berta Karlik was the head of the institute, and I did not have the impression the two of them were really close.

Marietta Blau continually gave seminar talks for up to fifteen people. These were mostly held on the ground floor [of the physics institute at Boltzmanngasse 5] in a large room and sometimes in a small lecture hall on the second floor. Sometimes she also gave lectures at the Physical Society, as I remember. She was frequently asked to give lectures, but she often refused saying: "I don't feel all that well." But when she gave talks abroad, she was particularly proud and happy.

MY DOCTORAL THESIS WITH MARIETTA BLAU IN VIENNA: 1960–1964
Gerda Petkov, Vienna[12]

I was very much impressed by Marietta Blau. I think Professor Karlik introduced me to her and from then on we had a really good relationship.

We measured meson-proton collisions, particularly the angle between the tracks, and drew conclusions about meson energy from them. She also looked for other particles. The work was very fascinating. When scanning photographic plates, we were helped by others with all the evaluations. After all, one had to search for events on these plates for months.

When Hannelore Sexl and I worked on the evaluation of photographic plates which had been irradiated with high-energy protons at CERN, Blau persuaded us to take a look at the experimental site. It was in this way that we embarked on the trip from Vienna to Bern and Geneva crammed into a VW beetle all the way.

I still remember her appearance: She was small and slight and had enormous dark eyes and very strong glasses, which made her eyes appear even larger. Often she appeared rather helpless to me. Although she was the professor who supervised my thesis, we had more personal than scientific contact with one another. She was something like a surrogate mother to me since she extended invitations to me, together with my then fiancé, and helped me out in my private life.

She had problems with burns on her hand, which sometimes appeared quite serious. She mentioned that from time to time she had severe pain; at those times, she would lock herself in her room so as not to disturb anybody else. She would never disturb anybody else. Once she made a statement which shocked me. She had very poor eyesight and needed strong glasses. Sometimes she crossed the street as if she were blind. This always worried me. Then she used to say: "If somebody runs over me, he'll have more trouble than I will."

---

[1] Hanne Ellis-Lauda, interview with Reinhard Schlögl, Vienna, February 1999.
[2] Pierre Radvanyi, personal communication to Robert Rosner, October 20, 1999.
[3] Martin M. Block, personal communication to Robert Rosner, December 13, 1999.
[4] Seymour J. Lindenbaum, personal communication to Robert Rosner, February 20, 2000.
[5] G. Bernardini, E.T. Booth, S.J. Lindenbaum, *Phys. Rev.* 80 (1950) 905, *Phys. Rev.* 83 (1951) 669, *Phys. Rev.* 82 (1951) 307, *Phys. Rev.* 85 (1952) 826, *Phys. Rev.* 88 (1952) 1017.
[6] Arnold Perlmutter, personal communication to Robert Rosner, March 2000.
[7] Sylvan C. Bloch, personal communication to Robert Rosner, April 3, 2000.
[8] Leopold Halpern, interview with Reinhard Schlögl, Vienna, September 1998.
[9] Brigitte Buschbeck, interview with Reinhard Schlögl, Vienna, February 1999.
[10] Herbert Pietschmann, personal communication to Brigitte Strohmaier, January 11, 2001.
[11] Hannelore Sexl, interview with Reinhard Schlögl, Vienna, February 1999.
[12] Gerda Petkov, interview with Reinhard Schlögl, Vienna, February 1999.

MARIETTA BLAU – THE SCIENTIST

Marietta Blau's scientific work was devoted to the development and application of the photographic method of particle detection in nuclear and particle physics. What were the foundations she was able to build on?

PHOTOGRAPHY AND EARLY RESEARCH ON RADIOACTIVITY

The word "photography" literally translates as "writing with light," and indeed the method was invented for taking pictures with visible light in 1837. In photographic exposure, an image is produced by means of an optical system in a light-sensitive emulsion consisting of silver-bromide grains, gelatin, and glycerin on a carrier, originally a glass plate. The light transfers part of its energy to the emulsion and thus decomposes some of the silver-bromide molecules. An invisible latent image is thus generated, which a developer transforms into a visible one consisting of shades of black caused by the presence of elemental silver. Fixing removes the unexposed and undeveloped substances and in this way produces a light-resistant negative picture. The image is called a negative because the light areas appear black and the dark areas light. A second analogous imaging step turns the negative into a positive image, thereby reproducing the original degrees of brightness.

It was found that this method of imaging also works with electromagnetic radiation of wave lengths outside the visible range. When Wilhelm Conrad Röntgen discovered x-rays in 1895, one of the first effects of this very penetrating radiation (with wavelength shorter than visible light) he observed was photographic. Almost immediately, he used x-rays to take pictures of human hands.

It was quickly observed that x-rays cause emission of visible light (luminescence) in certain substances. Among other scientists, Henri Becquerel started investigations in Paris in 1896; he exposed various minerals, among them uranium salts, to light of high intensity and then placed them onto photographic plates. Surprisingly, all of the uranium compounds, and only these, revealed an optical density, but it had nothing to do with luminescence and resulted from radiation penetrating the plate wrapping that was impermeable to light (now known as β- and γ-radiation; α-rays are absorbed in the wrapping). The property of emitting such radiation spontaneously was soon named "radioactivity" by Marie Curie and has presented challenges for science ever since.

Marietta Blau summarized the fundamentals of atomic physics when she gave a popular description of the use of photography in atomic research in the 1930s:

> In the radioactive substances we can directly observe the creation and annihilation of elements and recognize that, within these atoms, processes must take place which lead to a rearrangement of the constituents of the atom into more stable forms and to the production of new atoms by the ejection of atomic particles. These particles reveal essential information on the structure of the atoms and were identified shortly after the discovery of radioactivity. They are 1) β-particles, which are fast-moving, negatively charged particles, whose mass is only the eighteen-hundredth part of that of a hydrogen atom and which carry the smallest charge occurring in nature, an elementary quantum, and 2) α-particles, whose mass is four times that of a hydrogen atom and which possess a positive charge of two elementary quanta. The particles are emitted from the decaying atom with considerable velocity, characteristic of the radioactive element from which they originate. Usually α- and β-radiation are accompanied by a wave radiation of high frequency, a kind of x-ray, which is called γ-radiation. Radioactive transformations in which the γ-ray emission corresponds to the rearrangement of the nuclear constituents and which proceed without ejection of corpuscles from the atomic interior are also known.[B26]

Although the use of emulsions in radioactivity dates back to its discovery, it was of no particular importance in the research on nuclear radiation for several decades: The ionization of air by nuclear radiation was measured with electroscopes; scintillation counting was also used (i.e., registering the tiny light flashes occurring when radioactive radiation hits zinc sulfide, ZnS).

Nevertheless, it must be considered a milestone when Kinoshita[1] and Reinganum[2] identified trajectories of α-particles in photographic emulsions in 1910/11. As each grain hit by an α-particle is rendered developable, rows of dark grains of silver mark the passage of α-particles through the emulsion layer. Thus, if an α-particle strikes the surface of a photographic plate perpendicularly, it generates just one intense black dot; but if one sets the experimental arrangement such that the particle strikes at an angle, a row of developed grains can be detected (Fig. 31). The length of a row is a measure for the range of the particles in the emulsion. (The range [*Reichweite*] is the distance the particle travels until it is stopped due to successive energy losses.) The range in the emulsion can be translated into a corresponding range in air, which is related to the particle energy. The fact that these black tracks, the rows of silver grains, are of high contrast (there exists no continuous gradation from gray to black), was later explained by the large amount of energy of a single particle and by its specific mechanisms of energy dissipation.

At the Vienna Radium Institute in 1912, Wilhelm Michl began to investigate the detection of α-radiation with photographic emulsions.[3] He explained the effect of single α-particles incident on photographic plates under a grazing angle and tried to

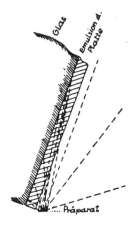

Fig. 31: Paths of α-particles in the emulsion of a plate[B26] (1934).

(Reprinted from *Photogr. Korresp.* 70, Suppl. 5 (1934) 31 with permission of Springer-Verlag, Heidelberg.)

understand the processes in the uppermost layer of the photographic plate. Michl was wounded in World War I and died in 1914. His studies were not carried on by others.

The beginning of nuclear physics came about when researchers bombarded matter with particles from radioactive decay and detected particles that were presumably emitted as a result. Much later, in 1934, Blau summarized:

> If α-particles pass through matter, they will be scattered diffusely to a larger or smaller extent depending on the material. Many α-particles undergo small deviations, while only a few are scattered through large angles, some even more than ninety degrees. If the α-particle only interferes with the electronic shell of the atom, the deviation from its direction is small due to the small mass of the electrons. If it interacts with the massive nucleus, it is reflected like a billiard ball on a wall at the corresponding angle. As the diameter of the nucleus is only one-$100,000^{th}$ to one-$10,000^{th}$ of the entire atom with its electron shells, scattering at large angles is much rarer because of the reduced probability.
>
> If the α-particles pass through a medium consisting of atoms whose nuclear mass is comparable to or even smaller than that of α-particles, the nuclei struck experience an appreciable momentum transfer. In fact, Marsden in Cambridge performed such experiments in 1914:[4] α-particles directed toward material containing hydrogen (paraffin) set single hydrogen nuclei in fast motion (Fig. 32). These nuclei were registered by observing scintillations on a zinc-sulfide screen placed in the path of the radiation.[B26]

The hydrogen nuclei set in motion by the recoil from α-particle scattering were called "natural H-particles."

However, there are other processes. In 1919 Ernest Rutherford, at that time in Manchester, had observed that a very small number of fast-moving particles result from collisions of α-particles with nitrogen gas.[5] These particles, which were identified as hydrogen nuclei, were called "H-rays"; in today's terminology, they are protons. More detailed studies led to the conclusion that the protons are knocked out of the nitrogen nuclei by

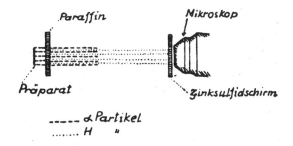

Fig. 32: Experimental setup for investigating natural H-particles[B26] (1934). (Reprinted from *Photogr. Korresp.* 70, Suppl. 5 (1934) 31 with permission of Springer-Verlag, Heidelberg.)

the colliding α-particles, i.e., the nitrogen nuclei disintegrate. In scientific notation, this reaction reads: $^{14}_{7}N\,(^{4}_{2}\alpha,\,^{1}_{1}p)\,^{17}_{8}O$. (The first symbol, $^{14}_{7}N$, nitrogen, indicates the target nucleus; the second, $^{4}_{2}\alpha$, the projectile; the third, $^{1}_{1}p$, the ejectile, i.e., the particle emitted from the composite system; and the fourth, $^{17}_{8}O$, oxygen, the residual nucleus. The superscripts denote the nuclear mass numbers of the particles, the subscripts the nuclear charge numbers. Frequently, the short form $^{14}N(\alpha,p)^{17}O$ is used.)

Following these experiments, the susceptibility to disintegration (disintegrability or *Zertrümmerbarkeit*) of other elements was investigated in laboratories all over Europe, and a considerable number of other elements were found to disintegrate under α-particle bombardment. In the process of disintegration, the α-particle can either remain in the nucleus, whereby the atom transforms into another with the charge number increased by one: the (α,p) reaction, or the α-particle does not lose its entire energy and leaves the nucleus with the result that two particles are ejected, and the charge number of the nucleus is decreased by one: the (α,pα) reaction.

The emitted fast protons, whose range in the gas exceeds that of the incident α-particles by a large degree, were registered by scintillation counting. Thus, the ZnS screen was positioned well beyond the range of the α-particles, and the experimenter looked at it through a microscope, waited for the occurrence of each of these rare and randomly distributed tiny flashes of light, and counted them (sometimes only a few events per hour). The procedure was tedious and easily subject to error.

Rutherford continued his investigations in Cambridge together with James Chadwick.[6, 7] Besides the use of scintillation counting, a more advanced instrument, Wilson's cloud chamber, was also available for studying nuclear reactions. Cloud chambers (pioneered by Patrick Blackett[8]) make the paths of charged particles with high velocity visible. The particle track is formed when a supersaturated vapor condenses because moving ions function as condensation nuclei. The droplets formed on them mark the particle's trajectory, allowing it to become visible and to be photographed.

MARIETTA BLAU'S INITIAL INVESTIGATIONS

At the Institut für Radiumforschung in Vienna in 1924, Hans Pettersson and Gerhard Kirsch attacked the problem of artificial nuclear disintegration, in German called *Atomzertrümmerungen* (smashing of atoms), using at first the scintillation method for particle detection.[9–11]

Pettersson (see p. 106) and Kirsch (see p. 106) soon considered it desirable to extend the methods for observing fast protons because both established techniques had severe deficiencies: the visual counting of scintillations was influenced by subjective factors, and experiments with Wilson's cloud chamber were difficult and time consuming. (Such a chamber was not constructed at the Vienna Radium Institute until 1926/27.) Therefore, Pettersson proposed to investigate "whether H-rays could be made amenable to objective observation through their photographic effect."[B6] (H-rays, H-particles, or H-nuclei were the terms used then for fast protons.) It was clear that if photographs of particle paths could be taken, the tracks would remain stored on the plates and enable subsequent analysis, in contrast to the quickly vanishing scintillations.

Marietta Blau presumably took up this question since she had previously dealt with the photographic effect of ionizing radiation. When she was employed at the Institut für Physikalische Grundlagen der Medizin in Frankfurt/Main in 1922, she studied the systematics of the photographic effects of x-rays and compared them to those of visible light in a publication with Kamillo Altenburger.[B2] The finding was that a single hit of an x-ray photon on a silver-bromide grain renders a grain of silver after development, whereas the threshold value in the gradation curves of visible light is explained by the assumption that a silver-bromide grain or a molecule complex needs to be hit by light photons more than once in order to be influenced effectively, as if it were excited in a multi-step process to a final state yielding a silver grain when developed. The intermediate states correspond to a loosening of the molecular structure and are not observable macroscopically.[B2]

There was great interest in the effect and registration of x-rays for medical applications, and in a second paper Blau published together with Altenburger,[B3] the authors point to the feasibility of applying the photographic method, i.e., the quantitative evaluation of exposed films, to determine the angular distribution of scattered x-rays.

FIRST PHOTOGRAPHIC DETECTION OF H-RAYS

When Marietta Blau started her photographic investigations[B5, B6] in Vienna in 1924, she mainly made use of fast protons ejected by nuclear collisions from substances containing hydrogen hit by α-particles of radioactive origin. Blau produced the sources of fast protons in such a way that α-radiating polonium was deposited electrolytically on small sheets of gold or platinum. On top of the polonium were placed thin layers of paraffin, in which the α-particles collided with the hydrogen nuclei (protons), thereby ejecting them, as explained above. In order to prevent a tremendous number of α-particles from reaching the photographic plates, the paraffin was covered with an absorption foil of copper or a combination of mica and copper, which does not keep the protons from leaving the paraffin.

After long and tedious studies, Blau succeeded in detecting natural H-particles. In the photographic layer, they act much the same as α-particles: they produce sequences of silver grains at grazing incidence. The detection of fast protons turned out to be much more difficult, however, particularly since even strong α-sources yielded only a small number of protons and since the photographic emulsions were also sensitive to the β- and γ-radiations emitted by the radioactive preparations as well as to light from luminescence and the laboratory environment. Therefore, the applied plates and developer types had to be chosen much more carefully than in the case of detecting α-particles.

Blau experimented with various types of photographic plates, among them ones that had successfully been used by Wilhelm Michl for registering tracks of α-particles. Soon Blau pointed to factors that determine the suitability of photographic plates for the detection of particle tracks: fine grain (i.e., small size of the silver-bromide crystals) and low sensitivity to competing radiation such as red light and β- and γ-rays. Besides the properties of the emulsions, the geometry also had to be set up appropriately. Because the emulsion layers were quite thin (200 μm maximum), the total track length could only be recorded if the protons passed the emulsion approximately parallel to the layer surface. Therefore, the protons from the source described above were selected by an aperture to produce a grazing incidence on the plates. Good results in registering proton tracks in emulsions were achieved with an angle of incidence of about 30°. Experiments replacing paraffin with soot yielded a significantly less silver-grain formation; thus, Blau made sure that the majority of the tracks obtained when paraffin was used indeed came from the H-rays.

Blau also carried out experiments to detect protons from the disintegration of aluminum by α-particles, i.e., the nuclear reaction $^{27}Al(\alpha,p)^{30}Si$ that had already been observed by other methods. The polonium on the flat metal backing was now covered with aluminum foil instead of paraffin layers. She found rows of silver grains in the expected direction and concluded that the reaction could be induced by α-particles of less than 4 cm range in air.[B6]

Even in these first studies with slow protons, Blau had observed that the density of silver grains in α-particle tracks is higher than that in proton tracks and concluded that the latter particle type has a lower ionizing power.[B5, B6] This finding was quite critical. Blau opened the gate to the field of particle identification because she could now distinguish tracks due to α-particles from tracks due to protons, as well as those from background events.

In 1928 she was able to make quantitative statements on the distance between the silver grains in particle tracks and compared the values for α-particles and protons. (Fig. 33 shows a proton track published by Blau.) This progress was made possible by the use of much smaller proton sources (a wire instead of a sheet), a varied geometry (angle of incidence 10–17°), and new photographic material.[B8, B9] The distances in α-particle tracks were significantly smaller than in proton tracks. Also, the higher the proton energy, the larger the distances were between the grains.

By this time, all evaluations of proton tracks in emulsions had established that the range in air and, applying the so-called range-energy relation, proton energy could be derived with sufficient accuracy from the track length.

Fig. 33: Track of a proton produced by a polonium α-particle colliding with hydrogen[B9] (1927).

(Reprinted from *Z. Phys.* 48 (1928) 751 with permission of Springer-Verlag, Heidelberg.)

SELECTION AND PRE-TREATMENT OF PHOTOGRAPHIC EMULSIONS

For experiments with α-particles, the choice of plates was not especially critical; fine-grain photographic plates rich in silver bromide turned out to be quite suitable.[B17, B21] On them, the α-particle encounters a large number of silver-bromide grains on its path through the emulsion; the track is pronounced and hence cannot be missed by microscopic investigation. Blau considered the detection of α-tracks sufficiently well established to allow her to start studying in detail the presumed fading of the latent image (i.e., whether and how fast the invisible image consisting of activated silver ions vanishes).[B16] Information about fading is necessary for quantitative radioactivity measurements if the photographic plates are stored for some time between exposure and development, and especially for low-level measurements in which it is desirable to choose long exposure times to register a significant number of events on a single plate. She found that the density produced by radiation decreases gradually, as much as 20% in sixty days. The effect is explained as a recrystallization process in the silver-bromide grains in which the radiation had caused lattice defects. These studies were later expanded by Johanna Lauda[12] and turned out to be most valuable in cosmic-ray research with exposures lasting several months.

In experiments with natural H-rays, the selection of plates is much more crucial and they are more difficult to carry out because some types of plates for H-particles yielded tracks with many fewer dots, and it was obvious that a row of dots could not be found for every incident H-particle. This means that not all the silver-bromide grains hit by an H-particle are sufficiently activated to be developed. Again, fine-grain plates proved to be quite effective for the detection of natural H-rays.

When Blau tried to use the same types of plates which had been successfully applied to natural H-rays to detect fast protons primarily from the disintegration of aluminum, she initially failed. But there were already methods being tested to reduce the sensitivity of photographic plates to certain types of radiation while leaving unchanged or even increasing their sensitivity to other types. For particle detection, this pre-treatment sought to decrease the sensitivity to light, β-, and γ-radiation without affecting the sensitivity to α- and H-radiation. The so-called desensitizers were various types of dyes. In her thesis,[13] Wambacher studied the influence of pinacryptol yellow on the sensitivity of photographic plates under Blau's supervision; together they investigated a large number of other substances.[B23] For the detection of fast protons, Blau and Wambacher chose pinacryptol yellow, with which they achieved a slight increase in grain density for α- and especially for H-particles. The image appeared distinctly on a clear background, thanks to the desensitizer.[B19] At first, the chemical effect of desensitizers could not be explained, even though they were used empirically. Blau and Wambacher, among others, attempted to provide an explanation.[B25, B27]

DETECTION OF NEUTRONS AND SPECTROSCOPY OF PROTONS FROM NUCLEAR REACTIONS

In 1930 Bothe and his collaborators at the University of Gießen in Germany showed that beryllium emits radiation of great penetrating power when bombarded by α-particles from polonium. Next, Irène Joliot-Curie and Frédéric Joliot at the Institut Curie in Paris showed that this radiation led to the ejection of protons from matter containing hydrogen. James Chadwick in Cambridge proved in 1931 that beryllium radiation consisted of electrically neutral particles of mass slightly larger than the proton mass.[14] In February 1932 Chadwick published an article, "Possible Existence of a Neutron," in the journal *Nature*.[15] Shortly thereafter, Werner Heisenberg formulated the proton-neutron model of the atomic nucleus with which we are familiar today. The constituents, proton and neutron, are collectively called nucleons, their number in a nucleus is called the mass number, and the nuclear charge number is equal to the number of protons.

Neutrons resulting from nuclear disintegrations leave the nuclei with tremendous velocity, and since they do not carry an electric charge, they do not interact with electric fields. They can therefore pass other atoms almost unhindered and can move in any material over long distances. They exert hardly any chemical or physical effects on their environment, which is why they cannot be observed directly and why they remained hidden for so long. But if a neutron collides with a substance containing hydrogen, it may hit a hydrogen nucleus and, much like a billard ball, set it in fast motion. In this way, the neutron manifests its own presence and kinetic energy.

Neutron experiments were carried out by placing an α-radiating preparation (polonium) close to beryllium. This arrangement acts as an intense neutron source. Close to the beryllium disk, a thin layer of paraffin rich in hydrogen was mounted. Collisions of α-particles with beryllium atoms led to the emission of neutrons which in turn accelerated hydrogen nuclei in the paraffin layer. These recoil protons could be detected by several techniques, photography being particularly suitable. It was not even necessary to insert a substance containing hydrogen between the beryllium and the photographic plate because the hydrogen content of the emulsion itself was sufficient to produce the "balls" thrust out by the neutrons. Because these balls were protons, they caused rows of dots on the photographic plates that were again pre-treated with pinacryptol yellow. From the lengths of the tracks, proton energies and consequently the energies of the neutrons could be estimated. It turned out that these energies (up to 9 MeV) were higher than assumed on the basis of other investigations at that time.[B19] (MeV is an energy unit used in nuclear and particle physics.)

Blau continued her experiments on neutron detection via recoil protons when she worked for several months in 1933 at the Paris Radium Institute upon invitation from Marie Curie. She analyzed approximately 3000 proton tracks and found that the neutrons emitted from the polonium-beryllium source were grouped according to energy, with some of them being extremely high in energy.[B22] For several years, the photographic method remained the only one able to detect the fastest of these neutrons.

Applying the range-energy relation, Blau used the photographic method to determine the energy spectrum (i.e., the intensity distribution as a function of energy) of protons emitted from the reaction $^{27}\text{Al}(\alpha,p)^{30}\text{Si}$. She began these experiments in Paris[B22] and continued them together with Wambacher in Vienna.[B24] An example of such a proton spectrum is displayed in Fig. 34; the maxima and minima are caused by the fact that the incident α-particles vary in energy. These investigations were continued in the research for doctoral theses by Elvira Steppan[16] and Otto Merhaut[17] that Blau supervised.

Whereas in these studies the track lengths of the protons were used as a measure of their energy, Blau and Wambacher also investigated the relationship between grain density (the number of silver grains per cm of track) and proton energy within given tracks.[B23] In accordance with Blau's earlier (1928) observations

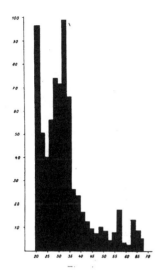

Fig. 34: Length spectrum of proton tracks[B24] from the reaction $^{27}Al(\alpha,p)^{30}Si$ (1934).

mentioned above, they measured an increase of grain density towards the end of the paths where the proton energy becomes less and less. In α-particle tracks, no such energy dependence was observed.

With their extensive experience in evaluating proton tracks, the two women later published a paper[B34] on the application of the photographic method to neutron spectroscopy (i.e., the distribution of neutrons with respect to energy) and to studies of neutron-proton scattering.

COSMIC RAYS: FAST PROTONS AND SPALLATION STARS

As soon as the photographic method was established as a tool for the detection of charged particles, another field of application opened up, namely the exploration of cosmic rays. Marietta Blau and her collaborator Hertha Wambacher turned to this field in 1932, twenty years after another scientist at the Vienna Radium Institute, Viktor Hess, had discovered cosmic rays.

♦ The fact that the air of the earth's atmosphere permanently possesses a small electrical conductivity had puzzled physicists since the end of the eighteenth century. Coulomb had observed that a charged sphere suspended on an isolating thread discharged slowly, but this phenomenon could not be explained by a leakage through the suspension. At the beginning of the twentieth century, when the presence of charged particles of radioactive origin was already known, the atmosphere's conductivity was thought to be due to ionization by energetic radiation, possibly from radioactive material in the earth's crust. In order to test whether the ionization decreased with increasing distance from the surface, Viktor Hess (see p. 107) measured the ionization with electroscopes aboard balloon flights originating in Vienna in 1911 and 1912. For his flights, he used free balloons filled with helium or hydrogen, which means they were subject to both thermal and up-and-down currents; the only means of control was releasing gas and reducing ballast. The flights carried him as far as Bohemia, and he reached a maximum altitude of over 5 km, the limit the human organism can safely withstand in an open gondola. At this altitude, he was able to measure beyond doubt an ionization approximately four times that on the ground. Hess concluded that the ionization must be caused by the extraterrestrial radiation of cosmic rays. For his pioneering work, Hess was awarded the Nobel Prize in Physics in 1936. Later research revealed that primary cosmic rays consist of positively charged atomic nuclei, among which protons form the largest portion. Nuclei of all elements up to the heaviest are present as well; their intensity decreases rapidly as the nuclear charge number increases. Primary cosmic rays induce nuclear reactions when reaching the earth's atmosphere. Short-lived particles (mesons, hyperons) as well as γ-rays, electrons, and positrons are produced in these reactions. As further interaction occurs, particle cascades and showers take place. ♦

In a few cases, heavy particles in cosmic rays had been registered with ionization chambers or Wilson's cloud chambers. With the latter instrument, it was difficult to take a picture at the exact time the particle passes the chamber. As a result, pictures were taken at regular intervals, and very few of them showed particle tracks. Blau and Wambacher recognized that the photographic method was superior to these techniques because the tracks corresponding to the sporadic occurrence of these particles could be accumulated in emulsions over long exposure times. Year-long trials were hampered by the fact that the photographic plates became foggy during long-term exposure. Evidently, the plates which had been treated with pinacryptol yellow were not suited to this task.

In 1937 Blau used a new type of plate (New Halftone Plate) that Ilford produced. The increased emulsion-layer thickness allowed for good proton detection. With this material, she approached Viktor Hess, who had installed a cosmic-ray observatory at an altitude of 2300 m on the summit of Hafelekar near Innsbruck (Tyrol). He allowed Blau and Wambacher to expose these new plates there. After an exposure of four months, they developed the plates and found two types of tracks: first there were numerous long tracks, which were interpreted as proton tracks. The lengths of the tracks indicated proton energies of a magnitude not previously observed. Many of the tracks corresponded to ranges in air of over a meter (i.e., 9 MeV) and up to 12 m (i.e., 40 MeV).[B35, B38] Since most of the tracks did not end within the emulsion, the observed track lengths were not a measure of the full energy of the particles. As a result, Blau and Wambacher attempted to determine the energy from the distances between the grains in the tracks. To be able to do so, they established a relationship between grain distance and particle energy by evaluating a large number of particle tracks. They calculated the energy of cosmic-ray protons to be predominantly around 12 MeV.

The second type of tracks observed in the Hafelekar exposures were short tracks radiating from a common origin. These strange, starlike tracks were called *Zertrümmerungssterne* (disintegration stars)[B35, B38] because they were interpreted to be the traces of disintegration processes (spallation) induced by cosmic-ray particles on the silver or bromine nuclei in the emulsion.[B40] (For an example, see Fig. 35.) The point from which the tracks led off indicated the location where the reaction took place, and the (three to twelve) prongs were the tracks of the fragments ejected. Blau and Wambacher initially found sixty such stars, of which they measured thirty-one precisely. The results were published in tabular form in their paper.[B38] Stars were also found in plates exposed at an even higher altitude, on the Jungfraujoch in Switzerland (3400 m).

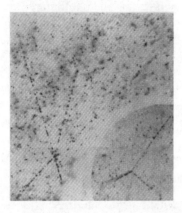

Fig. 35: Two disintegration stars[B38] (1937).

Control (reference) plates stored in Vienna were developed and compared to the Hafelekar plates: The former contained only about one-tenth of the number of proton tracks of the latter, a clear confirmation of the cosmic-ray origin of the protons. Blau and Wambacher assumed that the high-energy protons were recoil protons from neutrons passing through the plates' envelopes. These neutrons resulted from spallation processes outside the analysed emulsions.

THE *ANSCHLUSS* – AN ABRUPT END TO MARIETTA BLAU'S RESEARCH IN VIENNA

When the German troops occupied Austria in March 1938, Marietta Blau's scientific work was abruptly halted. After having achieved success with the publications on disintegration stars in 1937, she had become well known in the scientific community. A research visit to Oslo had been planned for some time. She left Vienna on March 12, 1938, one day after the Austrian government capitulated and never returned to Austria while it was under the control of the Nazis. Hertha Wambacher, her long-time co-worker, published further results on multiple disintegration of atomic nuclei by cosmic rays. In two of these reports[18, 19] Blau's contribution to this pioneering work is scarcely mentioned, and her decisive role in developing the photographic method in Vienna is not acknowledged. In a further paper Wambacher wrote together with Georg Stetter[20] and the summary[21] of a talk she gave, Blau is not even mentioned. Blau and Wambacher's work on spallation stars was continued in the investigations of Anton Widhalm.[22]

MARIETTA BLAU'S FINDINGS AS A STARTING POINT FOR NEW RESEARCH

Starting at the end of the 1930s, the working group founded by Cecil Powell (see p. 110) at the University of Bristol took on a leading role. In 1935 Powell decided to build a particle accelerator and work in the field of nuclear physics. The original plan was to study the scattering of fast neutrons by protons using a Wilson chamber for detection. In his autobiographical notes, he described how he learned from Blau and Wambacher's work and subsequently took over their approach:

> Around that time Walter Heitler, who had been in Bristol for a couple of years, drew our attention to the fact that Blau and Wambacher had successfully used photographic emulsions to detect particles from cosmic rays. Because the method had the virtue of being simple, we thought we could expose similar photographic plates on a mountain to see whether we could repeat the Viennese results ... The results were quite encouraging. We found "stars," certainly the result of the disintegration of nuclei, with up to ten charged particles emitted, protons and α-particles ... The modest results encouraged us to consider the use of photographic emulsions rather than the cloud chamber in our accelerator experiments to detect protons. Detecting neutrons (via protons) in a cloud chamber was tedious. In a typical experiment one had to take 30,000 pictures to obtain one particular neutron spectrum, which meant that the recording would have required six months ... We then brought some square centimeters of emulsion close

to the beryllium, which was bombarded with deuterons from the accelerator. With this emulsion we detected thousands of recoil protons and could determine the neutron spectrum with much higher accuracy than with the Wilson chamber, and the recording lasted only one or two days.[23]

Blau and Wambacher's discovery of the disintegration stars, the tracks of spallation processes induced by cosmic rays, was of interest not only to experimental nuclear physicists but also to theoreticians. Referring to their first short letter,[B36] Werner Heisenberg discussed processes occurring when high-energy protons or neutrons enter atomic nuclei.[24, 25] Heisenberg's particular interest in the experimental observation of spallation processes came from his prediction in 1936 that an energetic cosmic-ray particle hitting an atomic nucleus might cause an explosion-like production of a large number of particles.[26] Wolfgang Pauli's scientific correspondence[27] also indicates that the observation of the disintegration stars triggered discussions by important theoreticians such as Markus Fierz, Hans Bethe, Erich Bagge, and Pauli himself.

INTERLUDE IN OSLO

During Blau's stay at the Institute for Inorganic Chemistry of the University of Oslo, much of her time and energy was occupied with emigration matters for herself and her family and possibly for other physicists from the Radium Institute in Vienna.

In spite of all these concerns, she carried out a scientific investigation regarding α-radiation in tailings of zinc production.[B41] As early as in 1928, an α-radiation of 2 cm range in air had been found in these tailings[28] which could not be attributed to any of the then-known elements. Among them, α-particles of 1 cm range were observed in the zinc tailings, which later were attributed to samarium, as well as α-particles from the uranium-radium series and occasionally from the thorium series. The 2-cm α-particles always occurred in the presence of samarium. Studies on the existence and origin of this α-radiation had been performed at the Radium Institute in Vienna by Schintlmeister[29] in

1935/36. He had excluded various elements as emitters but made only vague assumptions as to which element could be the source of this radiation.

Blau intended to make a comprehensive measurement of ranges of α-particles from zinc tailings with the photographic method. The publication which resulted from her stay in Oslo[B41] describes experiments for which the photographic plates had been impregnated with solutions of samarium and uranium. In this way, these nuclides were distributed in the emulsions and particles emitted in their decay left tracks. The paper starts with a disclaimer stating that the study had been conceived of as a preliminary experiment for range measurement but was being published by itself "because the main experiments have been interrupted for some time now due to external factors" (evidently her plan to leave for Mexico). Blau's results consist of assumptions about which element chemically related to samarium might be the emitter of the 2-cm α-particles.

At the Vienna Radium Institute, the studies of this α-radiation and its possible emitter was continued through World War II, when the search for new elements was deemed important for military purposes, and several classified papers were dedicated to this topic. Although Blau's autobiographical notes written in the 1960s[30] indicate that the emitter was identified, the problem remained unsolved as late as 1971: Robert V. Gentry,[31] who had investigated unknown α-radioactivity and interviewed persons at the Vienna Radium Institute and the University of Mainz, could not make definitive statements regarding the 2-cm α-radiator.

BLAU'S PUBLICATIONS IN MEXICO

Einstein's reasoning for recommending a position for Blau in Mexico City had been that she was an expert at an inexpensive experimental technique and, therefore, would be able to spark scientific work even in a place where little equipment was available. Indeed, Blau had no apparatus at her disposal at the Escuela Superior de Ingeniería Mecánica y Eléctrica (ESIME). Her first publication in Mexico,[B42] which was also the first she wrote in Spanish, dealt once more with the 2-cm α-radiation. She presented an extensive evaluation of the results she had obtained in Oslo and speculated further as to which elements were the potential emitters of this radiation.

Blau's further publications in Mexico deal with areas with which she was concerned as a member of the Comisión Impulsora y Coordinatora de la Investigación Científica (CICIC). She was the head of the radioactivity laboratory of CICIC. The experiments mentioned in her papers were performed with equipment she had acquired or even constructed herself with the help of an engineer.

A paper on solar radiation in Mexico[B44] points to the influence of sunlight on problems in human medicine and biology, but also in geography, zoology, botany, and ecology, as well as in technology and pure science. Since the energy and spectral distribution of the insolation (i.e., solar irradiation) depend on geographical and climatological conditions, one can best determine these quantities by measuring them on a daily basis over several years under different atmospheric conditions with photoelectric cells or bolometers. Lacking the ability to conduct such long-term measurement, Blau chose a theoretical method for obtaining data on solar irradiation. Starting from the solar constant, which describes the solar power incident vertically on a unit area above the atmosphere, Blau arrived at a spectral distribution and a (seasonally dependent) energy flux of the solar radiation at the earth's surface calculated for Mexico with its low latitude and high altitude. She considered the effects of the scattering of sunlight on air molecules, water vapor, and dust particles, as well as absorption in the atmosphere along with geometrical effects. She also discussed the role of the ozone content, with ozone being both produced and destroyed by ultraviolet (UV) radiation of different wavelengths. Therefore, the quantity of ozone influences the amount of UV radiation present in the sunlight reaching the earth's surface. She emphasized the significance of UV radiation for biology and preventive medicine. Blau stated that UV light as a trigger for the formation of vitamin D is an important element in the prevention of rachitis and that it is a bactericide, preventing contagious diseases. Blau traced the reduced occurrence of illnesses in Mexico to the intense UV component in its sunlight and, with regard to the therapeutic effect of UV radiation, found Mexico's plateau superior to

Davos, a famous high-altitude health resort in Switzerland. All these findings clearly focused on public health issues in Mexico.

In further papers Blau published in Mexico, she described investigations of the earth's radioactivity. In the first[B43] of a series of publications Blau discussed the origin and localization of helium gas. In the ranking of the elements by their abundance in the cosmos, helium holds second place. As one of its production processes Blau explained its formation from a primitive mixture of protons and neutrons at the beginning of the universe. When helium is formed from two protons and two neutrons, a large amount of energy is released, and this heat played an important role in the evolution of the universe. Blau also mentioned the presence of helium in crystals of beryllium which cannot be traced to the influence of radioactivity. As an explanation, Blau suspected that the beryllium in these crystals can disintegrate through the action of gamma-rays into two helium nuclei and a neutron. She assumed the source of the γ-radiation to be cosmic rays rather than radioactive substances in the vicinity of the crystals.

But the most important production mechanism of helium is radioactive decay. Since the α-particles emitted by a large number of the members of the radioactive series are doubly charged helium nuclei, α-decay is related to the production of helium. From the total quantity of uranium and thorium in the earth's surface layer, millions of cubic meters of helium are produced each year which does not remain completely in the earth's crust but is released into the atmosphere at a rate of 30–60%. On earth, helium is found in two types of locations: in spring water, from which a large fraction escapes, and in natural gases (hydrocarburades) which contain helium and are associated with deposits of petroleum. These gases are of major importance for the commercial production of helium: although their helium content is low, the yield of the gases is very high. Regarding the joint existence of helium and petroleum, there seems to be a relation between them which also establishes a relation between petroleum and radioactivity. Apparently, certain organisms and bacteria which are responsible for the production of petroleum develop in the vicinity of old granite massifs and accumulate radium and uranium. As a consequence, helium is present in the oil deposits.

Another of Blau's studies which combined the fields of geology, physics, and chemistry sought to estimate the earth's thermic state,[B48] which is of great scientific importance for the understanding of earthquakes, volcanic activity, etc. The earth's surface temperature is determined by the equilibrium established between the heat it receives from the sun, the heat produced by radioactive substances in the earth's matter, and the heat the earth radiates into cold space. Assuming that in the beginning the earth was a heated mass, its process of cooling off to its current surface temperature was accompanied by solidification of the layers outside the earth's nucleus. Since then, the surface temperature has remained essentially unchanged over millions of years.

In order to calculate the integral effect radioactive substances contribute to the earth's thermal state, one must know their quantity and distribution. Existing experimental data on heat production from different types of rocks differed widely at the time of Blau's study. However, even using the lowest values, she estimated a temperature increase of thousands of degrees over the time span since solidification, if the radioactive substances are assumed to be equally distributed. Such an increase, however, would have made the solid crust melt and is therefore not plausible. Blau concluded that radioactive elements exist almost exclusively in a surface layer of 16–20 km. Investigation of thermal conditions depends on information about distribution of radioactive substances in the earth's interior. It may well be that these substances are distributed unevenly, that the granitic layer contains more of them than the basaltic layer and the nucleus. The estimated temperatures for the layer containing radioactive material depend on heat conductivity but are well below 1400°C. This means that this layer remains completely solid, in agreement with the evidence.

In order to study the distribution of radioactive substances in Mexico, a program to make the necessary measurements was initiated in the radioactivity laboratory of the CICIC. In her papers about this project,[B45, B46] Blau stated that the measurements served to establish a systematic classification of the radioactivity of minerals and waters within the Mexican Republic. In addition to the study of radioactive substances in minerals related to general geophysical questions, she also investigated radioactive spring waters and minerals relevant to medical and industrial purposes. She also studied the presence of radium and its progeny in crude oil as well as in water and sand containing oil.

Among the minerals which were suspected to be radioactive, the most important was uraninite, found in several of the mines at Placer de Guadalupe in the state of Chihuahua. Blau performed measurements on uranium oxide and radium and found considerable traces. Since the geological and mineralogical conditions suggested that radioactive minerals could be found in large quantity throughout the entire area, Blau suggested that it would be worthwhile to explore mines that were largely abandoned and in ruins because they potentially contained mineral treasures of immense value. In so-called non-radioactive minerals forming the earth's crust, radioactivity was measured to determine the distribution of radioactive substances in the earth, in particular, as a function of depth. Blau measured the radioactivity of samples from lava masses from volcanic eruptions. This field was considered particularly interesting in Mexico because of the numerous volcanoes and the large masses of lava.

Blau's paper on the measurement of small ionization currents[B47] treats a methodological question related to these investigations. The measurement of small currents as low as 10^{-10} A was possible with galvanometers and as low as 10^{-13} A with amplifiers. For even lower currents, more potent amplifiers, the so-called valve electrometers, were required. As these instruments were delicate and costly, Blau described the measurement of such small currents by use of conventional electrometers, which yielded reliable results only if their insulation was perfect. At that time, insulators were often made from melted quartz which was difficult to work with and whose insulating capability depended on the degree of humidity of the air, or from natural or artificial amber. Blau pointed to the fact that in practice there is a current through the insulator towards earth in addition to the current charging the electrometer. She recommended a procedure to identify a change in the resistance of the insulating parts without having to determine the value experimentally each time.

MOVE TO THE UNITED STATES

When Marietta Blau came to New York in 1944, she at first found herself at the periphery of the American research establishment. First she went to work for the International Rare Metals Refinery and later the Canadian Radium and Uranium Corporation. Perhaps the work she had done in Mexico on radioactive minerals helped to qualify her for these positions. Presumably, the positions in scientific research she might have preferred were not available to her as a non-U.S. citizen.

What is clear is that her pent-up frustration at having been relegated to scientifically remote Mexico now led to an explosion of creative activity.

THE FIRST SCINTILLATION COUNTER

Blau's first published paper after her arrival in the United States appeared in 1945[B49] and deals with the use of the photomultiplier tube to measure radioactivity. This was a setup for measuring the light intensity of scintillations which radiation (mainly from α-active polonium and radium samples) produces on a scintillating target (a zinc-blende [ZnS] screen). In the discussion, various applications of photomultiplier tubes for measuring radioactivity are presented. As mentioned earlier, the scintillation technique had been the first one used in detecting particles from artificial nuclear reactions. The drawbacks of having human observers count the light flashes were well known to Blau and had been the incentive for developing the photographic detection method between 1924 and 1926. But at the same time, Berta Karlik had investigated the scintillation method in her thesis and found that visual observation could be replaced by a photoelectric cell combined with an electrometer that would circumvent these drawbacks. Now Blau replaced the photoelectric cell with a photomultiplier tube, i.e., an electronic device consisting of a photocathode plus a series of electrodes (called dynodes), each of which is more positively charged than the one before it. In this cascade, one primary electron produces several secondary electrons at the surface of each dynode. Such an arrangement can multiply the tiny current emitted at the photocathode by many hundreds of millions.

The device was actually used as a dosimeter, that is, as a device for detecting integral radiation, but it must be regarded as the first rudimentary scintillation counter capable of resolving single rays. Electronic devices for amplification of individual pulses from radioactivity detectors were not available in Blau's laboratory. Therefore, in her publication, she reports on the use of conventional devices (milliammeter, microammeter, galvanometer) for measuring the current from the multiplier tube.

Further developments using liquid, plastic, and crystal scintillators soon made the scintillation counter a preeminent detector in nuclear and particle physics. In recent years, scintillation counting techniques have found a wide variety of important applications in biology, chemistry, geology, medicine, atmospheric science, and industry. One can trace the evolution of the modern scintillation counter using photomultiplier tubes from the very modest device produced by Blau to its present ubiquitous application across fields in science and technology.

It is unfortunate that Blau did not pursue the development of scintillation counter techniques further. Presumably the explanation is that because she was working for profit-making companies, which aimed at developing and selling radiation sources based on natural radioelements, she was not at liberty to pursue her own inclinations but had to follow the directives of management.

RESEARCH ON RADIOACTIVITY

During the years that followed, Blau carried out a number of projects[B49–B55] involving the measurement of radioactivity. Given that at this time her employers were mining companies, it is not surprising that her work involved the study of devices and procedures utilizing radioactivity. This was certainly true of her work on the scintillation counter described in the previous section,[B49] which aimed mainly at determining the activity of strong polonium sources. In her paper on that topic Blau also suggested the use of the zinc-blende screen as a secondary standard for calibrating the intensity of any light source.

Blau also examined the use of radioactive preparations (mainly of radium and polonium) to produce light from fluorescent materials (the phosphors).[B50] For calibration applications, the intensity of the light should be kept constant. However, fluorescent material, such as that used for watch and instrument dials, when combined with radioactive material, changes its luminescent emission because the continual bombardment by α-particles alters the crystal structure of the phosphor. But separation of the radioactive source from the fluorescent screen limits the irradiation of the screen to the relatively short intervals of actual use of the light source during which the light output remains relatively constant. A similar reduction of the irradiation can be achieved by the insertion of absorbing aluminum foil or variation of the gas pressure.

The radioactive standards for light could be very useful for colorimetric measurements and similar purposes. They have the advantage of being easier to control than ordinary standards for light because they do not involve electric currents that must be kept constant. Furthermore, in the case of radioactive light standards, the light output can be varied by varying the intensity of the radioactive preparation or the absorption of the radiation without changing the light color, whereas in the case of non-radioactive light standards, an increase in current or absorption influences the spectral distribution of the source.

In collaboration with the research director of the spa in Saratoga Springs, Blau and her co-workers addressed important issues in medicine and cancer therapy. They described a method[B51] of loading a chemical which can receive additional atoms, colloidal gamma ferric oxide (Fe_2O_3), with radon progeny (RaA [^{218}Po], RaB [^{214}Pb], and RaC [^{214}Bi]). Injected directly into the blood stream, this substance is ingested by phagocytes, a type of white blood cell. These cells are distributed throughout the body but are highly concentrated in the liver, spleen, and bone marrow. It has been observed that they can be influenced and their antibody effects stimulated if the colloidal γ ferric-oxide particles are combined with some therapeutically reactive material, such as certain radioactive substances that cause a high local dose of radiation in the cells. Blau and her collaborators also suggested that the inner radiation effect of activated ferric oxide might reduce the occurrence and growth of leukemia or tumors.

Another field in which Blau and her co-workers conducted research was in industrial applications of naturally radioactive elements. For example, they introduced a device for measuring surface areas[B52] which is based on extended flat polonium sources with highly uniform distribution of activity. A radiation-absorbing material between the source and an ionization chamber reduces the chamber current proportionally to the surface of the absorber. The experimental apparatus was proposed for measuring irregular areas as well as porosity and open areas of meshes.

Polonium could also be used as a reference for measuring the activity of α-active protactinium in very thick samples in ionization chambers.[B53] When the absorption of α-particles in the sample layers has to be determined, the addition of a known quantity of polonium to thick (α-saturated) samples makes this measurement possible.

Moreover, the research included technical details of representative new devices based on radioactive sources:[B54] electrostatic voltmeters (discharge via ionization of air); resistors (ionization of gases); measurement of static electricity for charge monitoring; light sources (including temperature display based on the decrease of radio luminescence of certain substances with increasing temperature); a thickness gauge for moving sheets, e.g., in rolling mills, measuring the penetrating β-radiation. (The authors suggested the use of radium as a β-source covered with an extremely thin metal layer to suppress the efflux of radon and α-particles.)

Further metrological applications included liquid level detection, e.g., with an α-source on a float, and displacement detection by applying the radioactive material to a moving part of a system and measuring the varying ionization current, e.g., in microbalance arms. In this paper, Blau and her collaborators also mentioned for which of the applications they had filed patents. Finally, the authors emphasized that the devices discussed in this paper were merely representative examples of a wide range of industrial applications whose number was likely to increase.

In her last publication[B55] on her work for industrial companies, Blau described her attempts to develop a practical device to measure the ionizing power of β-rays in condensed materials.

BLAU'S RETURN TO THE PHOTOGRAPHIC METHOD IN 1948

In 1948 Marietta Blau was employed as a research physicist at Columbia University as the result of two factors which complemented one another. Firstly, Blau had, of course, followed developments in the use of photographic emulsions and their application in studying the phenomena of particle physics that led to the discovery of the π-meson (pion). She longed for a chance to return to her primary research interest in the emulsion technique.

♦ The discovery of the π-meson (pion) was the central event in nuclear physics after World War II. This discovery not only led to finding numerous other particles but also resolved existing contradictions. In 1935 Hideki Yukawa, the great Japanese theorist, attributed the interaction between nucleons (protons and neutrons), i.e., the nuclear forces, to the exchange of a massive field quantum, estimating its mass at about 300 electron masses. Two years later, C.D. Anderson observed new charged particles of about 200 electron masses in cloud chambers exposed to cosmic rays. The particle, called the μ-meson or muon, seemed to confirm Yukawa's prediction. However, the mass did not correspond well to the theory and, furthermore, the muon did not interact strongly in nuclear matter, as was expected of the Yukawa particle, but rather exhibited weak forces. The muon is actually a lepton but retains the name meson for historical reasons. The discovery of the pion in 1947 resolved the problem: It was suggested that in Anderson's observation two particles (a muon plus an undetected neutrino) were involved. The pion strongly interacted with nuclei, in agreement with the Yukawa prediction, and confirmed his interpretation of nuclear forces.

The π-meson was discovered by observing emulsion tracks which registered the decay of stopped pions into muons. (The emulsions were exposed at 5500 m altitude in the Bolivian Andes and at 2800 m altitude in the Pyrenees.) Grain densities and scattering measurements identified the primary and secondary particles as mesons and indicated the directions of their motion. ♦

Secondly, the newly established Atomic Energy Commission shifted its research program to high-energy physics as a result of the anticipated construction of particle accelerators, which had begun in 1947 in the United States. Hence, efficient working teams for the detection of high-energy particles and their reactions were required. Both the adaptation of the photographic method, which now proved to be of critical importance, and the construction of cloud chambers and bubble chambers, as well as setups of scintillation counters were needed. There were no experts on the emulsion technique at American universities or at the research laboratories connected with the Manhattan Project.

♦ The first stage in the construction of accelerators for high-energy physics in the United States in the years after World War II took place at several universities where cyclotrons yielding particles with energies of several hundred MeV were built. One of them, the Nevis Cyclotron at Columbia University, could not be housed on the campus in New York City and was therefore located in a newly erected block of houses near Irvington, N.Y. about twenty miles north of New York City. At the same time, high-energy accelerators for particles of several GeV energy were being developed and constructed: the Cosmotron for 3-GeV protons at the newly built Brookhaven Laboratory on Long Island, N.Y. and the Bevatron for 6-GeV protons at University of California, Berkeley. (GeV is the abbreviation for the energy unit Giga electron Volts; $1 \text{ GeV} = 10^9$ eV. Before 1960 the notation BeV [Billion electron Volts] was frequently used.) The Brookhaven Cosmotron was the first accelerator worldwide to deliver a proton beam of more than 1 GeV energy. It became operational in 1952, yielding at first 2.3-GeV protons and later on the intended energy of 3 GeV. As designated by the Atomic Energy Commission, the Brookhaven Laboratory was operated and administered by AUI – Associated Universities Incorporated. This company comprised nine universities in the northeast United States: Columbia, Cornell, Harvard, Johns Hopkins, MIT, Pennsylvania, Princeton, Rutgers, and Yale. The close cooperation of Columbia and Brookhaven explains the fact that Blau at first was employed at Columbia University and about two years later joined the Brookhaven staff. ♦

Blau's subsequent publications in the field of emulsions testify to her exceptional dedication and energy in her new employment. She appears to have thrown herself totally into the emulsion research from which she had been absent for ten years. Despite the fact that she was effectively an outsider in the American research establishment, her output during this time was prodigious. Methods of handling and developing emulsions[B56] and properties of particle tracks were the main subjects Blau and her new emulsion group at Columbia University investigated. Thus, Blau's renewed foray into the emulsion field after ten years of absence led to an important advance in working methodology and bears witness to her insight and meticulousness.

Observed grain density is one of the principal parameters for evaluating particle-track measurements in emulsions because it enables the identification of the species and/or energies of particles passing through the matter. Blau describes the development of theoretical formulas for the dependence of grain density (the number of developed silver-bromide grains per length of particle track) on particle species and energy. She compared her formulas containing only a few empirical parameters with the experimental results (for α-particles, protons, tritons, π-, and μ-mesons)[B57, B58] and found very good agreement.

It is quite striking that in her work on grain density, as well as in the previous research on emulsion development, Blau methodically reeducated herself on the techniques of experimental emulsions and acquainted herself with the principal technical developments since 1938.

As mentioned earlier, one of Blau's tasks at Columbia University was to instruct researchers on the techniques of using photographic emulsions at the Nevis Cyclotron, which was then under construction. Apparently, they exposed some emulsions in balloons for practice and encountered a strange event caused by high-energy cosmic rays in the emulsions.[B59] Since the "experimental particle zoo" was being assembled in those years, each event was accorded special attention. In this case, the event in the cosmic-ray star was interpreted as the capture of a τ-meson (now called the K-meson or kaon) by a bromine or silver nucleus in the emulsion.

♦ The discovery of the pion ushered in the great era of high-energy physics in the second half of the twentieth century. New particles were discovered which, due to their large number and the fact that they could not at first be classified in systematic order according to their nature, were referred to as the particle zoo, and new working methods came about thanks to the construction of high-energy accelerators and the development of metrology. The most important discoveries are listed below, indicating the use of photographic emulsions (em.) and particles of cosmic rays (c.r.) or particle accelerators (p.a.):

- Charged pions ($\pi^{\pm}$) 1947 (em., c.r.)
- Uncharged pion (π^{o}) 1950 (p.a.)
- Heavy mesons 1949 (em., c.r.). Before their masses were known exactly, the particles were characterized by their modes of decay and named τ-, K-, θ-, χ-particles. When mass detection resulted in the same value for all of them, the conclusion was drawn around 1953 that they were a single particle species (K-meson, kaon; $K^{\pm}$, K^{o}) that can undergo many transformations.
- Hyperons, i.e., particles heavier than nucleons (proton and neutron); 1954–1955. The nucleons and the hyperons together make up the baryon family of particles. Important hyperons are Λ^{o} (c.r.), $\Sigma^{\pm}$ (c.r.), Σ^{o} (p.a.), Ξ^{-} (c.r.), Ξ^{o} (p.a).

- The production of high-intensity meson and hyperon beams with particle accelerators starting in 1948 opened new important possibilities in particle physics.

The classification scheme of elementary particles which is generally accepted today was introduced[32, 33] in 1961–1962. Particles are characterized by several quantities and quantum numbers (mass, charge, spin, isospin, strangeness, parity). In this scheme, baryons (nucleons and hyperons) as well as mesons form families of eight members, so-called octets.

This classification scheme was given a theoretical ground in 1964 by the introduction of quarks as constituents of strongly interacting particles.[34, 35] It was postulated that there are three types of quarks and their antiparticles. Later, three more types of quarks were postulated to explain certain particle properties. Theory posits charges of $+2/3$ and $-1/3$ of the elementary charge (i.e., that of the electron). Baryons consist of three, mesons of two quarks. Quarks have not been seen in a free state, since they are confined to elementary particles by strong forces involving also gluons. Hence, they can be detected only indirectly, but their utility in bringing an order to hundreds of particle states has been of inestimable value. ♦

In another investigation[B63] on emulsions which were exposed in a balloon at an altitude of 28 km, Blau and her co-workers studied the effect of copper and lead absorbers of varying thickness on the rate of star and meson production.

In a methodical study,[B60] the measurement of slow neutrons with emulsions was investigated. Untreated β-sensitive emulsions, emulsions loaded with additional elements (in particular boron), and combinations of metal foils (especially indium) with emulsions were investigated. Here the additional matter served to generate charged particles in neutron-induced reactions. Since the detection efficiency depends upon neutron energy, experiments were performed with neutrons of three energy ranges: epithermal, thermal, and cold.

With the construction of a semi-automatic device for analyzing events in nuclear emulsions,[B62, B65] Marietta Blau, along with Robert Rudin and Seymour Lindenbaum, made a significant contribution to the photographic method. This work was a landmark which led not only to advances in analyzing emulsion tracks but also portended much later developments in the analysis of bubble-chamber, spark-chamber, and streamer-chamber photographs. When one considers the rudimentary level of computers and optical devices of that time, the capabilities of this device are remarkable and deserve admiration.

The instrument (see Fig. 36) was built around a microscope with a stage driven by synchronized motors in two dimensions. The accuracy of gears, feedscrews, etc., was such that dimensions could be measured with extremely high precision. The photographic plates could be moved in any direction at any desired speed (up to 25 µm per second). For convenience, the device was controlled by a steering wheel and the speed of scanning was controlled by a pedal. There was also a recording

Fig. 36: Device for the semi-automatic evaluation of tracks in emulsions[B65] (1950). (Reprinted from *Rev. Sci. Instrum.* 21 (1950) 978 with permission of the American Institute of Physics.)

chart that moved at a speed 2000 times that of the stage. The image on the plate was observed by the operator through the eyepiece and at the same time was projected on a small slit in front of the multiplier tube.

The measurements were made quite rapidly and easily. For example, a conservative estimate of the driving time for the grain-density record of a 2-mm track was about ten minutes. By observing electron tracks emanating from the main track, the system was adaptable for the measurement of tracks of particles with high (3 to 26) nuclear charge numbers.

MARIETTA BLAU AT BROOKHAVEN NATIONAL LABORATORY

After her move to Brookhaven National Laboratory (BNL) in 1950, Marietta Blau mainly published the results of her experiments with accelerators.

Both for understanding the interaction processes of high-energy particles with emulsions and for interpreting experiments using emulsions, it is significant whether the reactions occur on the (heavy) elements of silver bromide or on the (light) elements of gelatin. In order to study this problem, Blau and her collaborators used emulsions sandwiched between thin layers of pure gelatin.[B65, B67] Incoming particles produced in the Nevis Cyclotron of Columbia University were 300-MeV neutrons and 50 to 80-MeV π^+-mesons. The number and properties of the stars produced in the gelatin layers and in the emulsions were compared. When irradiating the emulsions with neutrons, the ratio of stars induced in the light elements to those induced in the heavy elements of the emulsion was determined to be 1:5. Furthermore, the researchers found that the mean number of particles per star in reactions with light nuclei was larger than in those with heavy nuclei. They also determined the ratio of the number of α-particles produced in stars to the number of protons produced. The stars (for an example, see Fig. 37) produced by π^+-mesons with energies of 50–80 MeV incident on emulsion sandwiches originate in the light nuclei of the emulsion at a rate of 24–30% that of those with heavy nuclei.

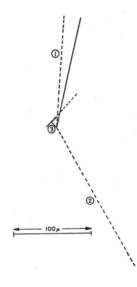

Fig. 37: Nuclear reaction induced by a pion (π^+) on a carbon or nitrogen nucleus[B68] (Nevis Cyclotron; 1953).

Tracks:
1: incident pion
2: emitted proton
3: emitted nucleus, presumably ^{8}Be
4: unidentified particle (proton, deuteron, or α-particle).

(Reprinted from *Phys. Rev.* 91 (1953) 949 with permission of the American Physical Society.)

Blau was also involved in work at BNL on the interactions of 500-MeV negative pions (π^--mesons) with emulsion nuclei. One important result was the observation of meson interactions with nuclei, producing additional mesons[B69] (see Fig. 38). While the authors give credit to other research groups for finding meson-production events in meson interactions at lower energy, it may be stated with considerable justification that Blau's was the first definitive report of additional-meson production by high-energy mesons. A further goal of these studies[B69–B71] was to find the dominant interaction of negative pions of such high energies, compared to mesons of lower energy. The experiments, however, showed a complex variety of processes. Inelastic scattering of mesons, i.e., star production with one or more mesons emitted, occurred more frequently at higher energies. There is also a small probability for the emission of two charged mesons and for the production of neutral pions (π°-mesons). Blau and her co-workers compared the results with those of cosmic-ray mesons, where the mean shower energy is 640 MeV. There appeared to be a considerable discrepancy in the results with shower mesons and with single artificial mesons, which the group was at a loss to explain in the early 1950s.

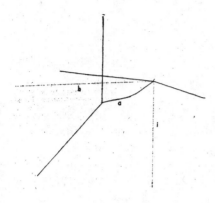

Fig. 38: Meson production in a meson-induced reaction[B69] (Brookhaven Cosmotron, 1953).

Tracks:
i: incident π^-
a, b: emitted mesons.

(Reprinted from *Phys. Rev.* 92 (1953) 516 with permission of the American Physical Society.)

Blau and her colleagues expanded the investigation of negative-pion interactions with emulsion nuclei to higher pion energies (750 MeV).[B72] They scanned a total of 133 meters of meson tracks under high magnification and found 322 interaction events. They analyzed the results according to the type of interaction, i.e., whether the pion energy is unchanged (elastic scattering) or whether additional-meson production occurs, mainly in scattering by free protons or protons at the periphery of the nucleus. In some events, charge exchange was observed, that is to say, the interaction of negative pion and proton yields either two mesons of opposite charge and a neutron, or two uncharged pions and a neutron.

Blau's last publication, in 1956, on research conducted at Brookhaven National Laboratory was on hyperfragments and K-mesons in stars produced by 3-GeV protons.[B73] (Hyperfragments consist of hyperons, i.e., unstable or metastable particles heavier than nucleons, bound to ordinary nuclei or fragments of nuclei. These types of aggregates were discovered in emulsions[36] in 1953 and then investigated at various laboratories, among them the Cosmotron laboratory at BNL.) A stack of emulsions consisting of twenty-four layers was exposed to 30,000 protons/cm^2 parallel to the layers. Of the 14,480 stars investigated, Blau found fourteen events which were believed to be spontaneous disintegrations of hyperfragments coming to rest in the emulsion. These decays yielded secondary stars. As far as the events could be analyzed, they were compatible with the decay of a hyperfragment, namely a Λ^0 hyperon bound to the nucleus. Analysis of the observed stars showed that different nuclear fragments carry the hyperon: H^4 (or H^3), He^5 and He^6, Li^6, Li^7, Li^8 and Li^9, Be^8 and Be^9, B^9 and B^{10}, $C^{10...13}$, $O^{16...}$. (The superscripts indicate the nucleon number plus one for the hyperon.)

The number of hyperfragments per star is one-tenth of one percent, which essentially corresponds to the results of other researchers. All hyperfragments originate in heavy elements, which are reponsible for 75% of all stars.

In summary, it is clear from Marietta Blau's research at BNL that she authoritatively entered into the mainstream of particle research despite the fact that she was not in charge of large research groups. In particular, she quantified the interaction of (then) high-energy pions, including the discovery of the first examples of additional pion production. She also contributed significantly at an early stage to the observations of hyperfragments. Although the improvement of statistics, i.e., the observation of a much larger number of events, came several years later with the use of hydrogen, deuterium, and helium bubble chambers, the course of this further research was paved by the results from the fourteen events using emulsions.

MARIETTA BLAU'S RESEARCH AT THE UNIVERSITY OF MIAMI

Marietta Blau came to the University of Miami at Coral Gables as an Associate Professor in 1956. She had generous funding from the Air Force Office of Scientific Research which she used for building up an emulsion laboratory. It was equipped with six or seven precision Leitz binocular microscopes with magnification up to 2000×, for which her co-workers had designed enlarged movable stages, built by local instrument makers.

When the laboratory at the University of Miami was established, Blau and her group began to scan a stack of emulsions that had been exposed to the π^--beam at the Brookhaven Cosmotron.[B75] Arnold Perlmutter took part in this study. The track length scanned was 381 meters, and the number of pion interactions found was 811. The authors separated interactions with free protons from those with protons bound in nuclei. Again, the events were analyzed according to interaction type, the main type (49 events) being interactions of the π^- on free or edge protons, with either no energy change (elastic scattering) or production of an additional neutral pion, or with two charged pions of opposite charge plus a neutron in the outgoing channel. A group of collisions was attributed to protons but could not be analyzed with regard to outgoing particles. About the same number of collisions occurred on nuclear-edge neutrons.

The results for fast-pion interactions on free protons with the production of an additional pion gave early evidence of the formation of a nuclear resonance, which in the quark model of Gell-Mann became known as the Δ-particle and which decays into a nucleon and a pion. This result was corroborated a few years later by others, with many more events.

During Blau's stay in Miami she vigorously attacked the problems of ionization in nuclear emulsions,[B76] which she had addressed earlier.[B57] During the course of this investigation carried out with three collaborators, measurements were made on known tracks (of protons, pions, and antiprotons) for the purpose of particle identification. It was found that a single parameter, namely the blob density, could be related to the probability of ionization over the entire energy range. The blob density is the number of silver-bromide single grains or clusters of grains per unit length of track. This quantity replaces counting individual silver-bromide grains and avoids the uncertainties which might occur as a result of closely spaced grains. The experimental apparatus was inspired by the earlier device designed by Blau and her collaborators[B65] for analyzing particle tracks in emulsions.

Another of Blau's projects concerned the analysis of emulsions exposed at the Berkeley Bevatron. These experiments were performed in 1957–1959 with particle beams consisting of K^+-mesons, antiprotons, and K^--mesons.[B78, B80] Blau and her colleagues observed a substantial number of K^--mesons stopped in the emulsion but, because of the large number of events accumulated by the European K^- Collaboration, restricted themselves to describing[B80] only some unusual interactions of hyperons, i.e., reactions of hyperons and hyperfragments produced when K^--mesons were absorbed on nuclei in the emulsion.

The authors also found several examples of stopping K^- interactions with two nucleons; the exotic decay, $\Sigma^+ \to p + \gamma$; and the decay of a hyperfragment, $\Sigma^- + p \to \Lambda^0 + n$. Finally, a sample of four scatterings by Σ^- or Σ^+ interactions was added to those found by other groups. From these data, the total mean free path of K^--mesons in an emulsion was calculated.

Clearly, this work, co-authored again by Arnold Perlmutter, was a significant contribution to the early understanding of hyperon interactions.

REVIEW ARTICLES AND OTHER WRITINGS BY MARIETTA BLAU

During her years in the United States, Marietta Blau published four more works in addition to the research articles described above. These include review papers and a retrospective on the discovery of cosmic-ray stars in photographic emulsions. These publications are of interest not only for their scientific content but also because they reveal Blau's efforts in a twofold manner. First, they reflect her attempts, dating back to the 1930s, to inform a wide circle of scientists about progress in her own field of research through review articles and secondly, her desire to establish a research relationship with the Vienna Radium Institute after the end of World War II.

A retrospective report on the discovery of cosmic-ray stars in emulsions[B61] was published in 1950 on the occasion of the fortieth anniversary of the opening of the Vienna Radium Institute. Blau gave a brief history of her work with photographic emulsions in the 1920s and 1930s, of her collaboration with H. Wambacher, and of the exposure of plates at Hafelekar. She described the stars found in the emulsion and explained why they were to be interpreted as products of incident cosmic rays. She mentioned the planned expansion of these experiments by exposing plates, partially covered by thin foils of various materials, at different altitudes and latitudes. The project was rendered impossible, however, because of political developments.

In the same year, Blau published a review article on the applicability of the photographic method in nuclear physics and cosmic-ray research in a publication (*Festschrift*) in honor of

Karl Przibram's seventieth birthday.[B64] (For biographical details on Karl Przibram, see p. 104.)

Blau wrote the third paper in this group in 1959 while she was at the University of Miami. It deals with ionization measurements in photographic emulsions[B74] and was published for Karl Przibram's eightieth birthday. This paper is based in part on "Studies of Ionization Parameters in Nuclear Emulsions,"[B76] discussed above, and on another series of articles[B79] she was writing at that time for the monograph *Methods of Experimental Physics*, edited by Luke C.L. Yuan and Chien-Shiung Wu. Blau's contributions to this work are a masterful discussion of the photographic emulsion technique in nuclear and particle physics by the person who is most responsible for its discovery and for much of its evolution and application. The main topics of Blau's articles include:

- questions that are clearly addressed to the concerns of emulsion practitioners, namely, on the sensitivity of various emulsions, processing techniques of nuclear emulsions, water content of emulsions, and optical equipment and microscopes
- the range of particles in nuclear emulsions, in particular range-energy relations, ionization measurement in emulsions, and various parameters characterizing the ionization produced
- determination of the charge of particles in photographic emulsions
- momentum measurement in nuclear emulsions
- detection and measurement of gamma-rays in photographic emulsions
- determination of particle masses, particularly from the range-energy relation and the average scattering angle.

Even with the passage of forty years, Blau's overview remains an important contribution to the literature on experimental particle physics. Had Blau been in better health, this exposition could have become a classic book on the emulsion technique in high-energy physics.

A Russian translation of the first part of Blau's contributions to *Methods of Experimental Physics* was published in 1963.[B81]

Marietta Blau's early work on the photographic method and its applications provided an essential contribution to the great discoveries made later with this technique. Due to her long absence from pioneering research between 1938 and 1944, she was no longer in a position to forge new roads after the war, even though she still accounted for significant advancements in this field. This was a period when the photographic method was being successfully applied in many laboratories. Emulsions were largely supplanted by bubble chambers in the 1960s, and then by spark chambers, later by streamer chambers, and finally by a whole new class of detectors.

Abbreviations used:
MIR   *Mitteilungen des Instituts für Radiumforschung*
Sondersammlung Special collection of the Österreichische Zentral-
Zentralbibl. Phys. bibliothek für Physik, Vienna.

[1] S. Kinoshita, "Photographic action of the α-particles," *Proc. Roy. Soc.* A83 (1910) 432.
[2] M. Reinganum, "Streuung und photographische Wirkung der α-Strahlen," *Phys. Z.* 12 (1911) 1076.
[3] Wilhelm Michl, "Über die Photographie der Bahnen einzelner α-Teilchen," *Sitzungsber. Kaiserl. Akad. Wiss. Wien, Math. Naturwiss. Kl. IIa* 121 (1912) 1431; Wilhelm Michl, "Zur photographischen Wirkung der α-Teilchen," *Sitzungsber. Kaiserl. Akad. Wiss. Wien, Math. Naturwiss. Kl. IIa* 123 (1914) 1955.
[4] Ernest Marsden, "The passage of α-particles through hydrogen," *Phil. Mag.* (Ser. 6) 27 (1914) 824.
[5] Ernest Rutherford, "Collision of α-particles with light atoms, IV. An anomalous effect in nitrogen," *Phil. Mag.* (Ser. 6) 37 (1919) 581; "Nuclear constitution of atoms," *Proc. Roy. Soc.* A 97 (1920) 374.
[6] Ernest Rutherford and James Chadwick, "Artificial disintegration of light elements," *Phil. Mag.* (Ser. 6) 42 (1921) 809; "The disintegration of elements by α-particles," *Phil. Mag.* (Ser. 6) 44 (1922) 417.
[7] Ernest Rutherford and James Chadwick, "Further experiments on the disintegration of elements," *Proc. Phys. Soc.* 36 (1924) 417.

[8] Patrick M.S. Blackett, "Ejection of protons from nitrogen nuclei – photographed by the Wilson method," *Proc. Roy. Soc.* A 107 (1925) 349.

[9] Gerhard Kirsch and Hans Pettersson, "Über die Atomzertrümmerung durch α-Partikeln," *Sitzungsber. Akad. Wiss. Wien, Math. Naturwiss. Kl. IIa* 132 (1923) 299; Gerhard Kirsch and Hans Pettersson, "Über die Atomzertrümmerung durch α-Partikeln II. Eine Methode zur Beobachtung der Atomtrümmer von kurzer Reichweite," *Sitzungsber. Akad. Wiss. Wien, Math. Naturwiss. Kl. IIa* 133 (1924) 235; Gerhard Kirsch, "Über die Atomzertrümmerung durch α-Strahlen IV. Abbau von Stickstoff und Sauerstoff. – Helium als Abbauprodukt," *Sitzungsber. Akad. Wiss. Wien, Math. Naturwiss. Kl. IIa* 133 (1924) 461.

[10] Gerhard Kirsch and Hans Pettersson, "Experiments on the Artificial Disintegration of Atoms," *Phil. Mag.* (Ser. 6) 47 (1924) 500.

[11] Hans Pettersson and Gerhard Kirsch, *Atomzertrümmerung, Verwandlung der Elemente durch Bestrahlung mit α-Teilchen* (Leipzig: Akademische Verlagsgesellschaft, 1926).

[12] Johanna Lauda, "Über das Abklingen des latenten Bildes auf der photographischen Platte," *Sitzungsber. Akad. Wiss. Wien, Math. Naturwiss. Kl. IIa* 145 (1936) 707.

[13] Hertha Wambacher, "Untersuchung der photographischen Wirkung radioaktiver Strahlungen auf mit Chromsäure und Pinakryptolgelb vorbehandelte Filme und Platten," *Sitzungsber. Akad. Wiss. Wien, Math. Naturwiss. Kl. IIa* 140 (1931) 271.

[14] The observed nuclear reaction, therefore, was $^9_4\text{Be}(^4_2\alpha, ^1_0\text{n})^{12}_6\text{O}$.

[15] James Chadwick, "On the possible existence of a neutron," *Nature* 129 (1932) 312.

[16] Elvira Steppan, "Das Problem der Zertrümmerung von Aluminium behandelt mit der photographischen Methode," *Sitzungsber. Akad. Wiss. Wien, Math. Naturwiss. Kl. IIa* 144 (1935) 455.

[17] Otto Merhaut, *Das Problem der Resonanzeindringung von α-Teilchen in den Aluminiumkern, behandelt mit der photographischen Methode*, Dissertation, University of Vienna (1938).

[18] Hertha Wambacher, "Mehrfachzertrümmerung von Atomkernen durch kosmische Strahlung; Ergebnisse aus 154 Zertrümmerungssternen in photographischen Platten," *Physikal. Z.* 39 (1938) 883; idem, *Z. techn. Physik* 19 (1938) 569.

[19] Hertha Wambacher, "Kernzertrümmerung duch Höhenstrahlung in der photographischen Emulsion," *Sitzungsber. Akad. Wiss. Wien, Math. Naturwiss. Kl. IIa* 149 (1940) 157.

[20] Georg Stetter and Hertha Wambacher, "Neuere Ergebnisse von Untersuchungen über die Mehrfachzertrümmerung von Atomkernen durch Höhenstrahlen," *Physikal. Z.* 40 (1939) 702.

[21] Hertha Wambacher, "Mehrfachzertrümmerung von Atomkernen durch kosmische Strahlung," *Angew. Chemie* 52 (1939) 117.

[22] Anton Widhalm, "Schwere Teilchen in der kosmischen Höhenstrahlung," *Z. Physik* 115 (1940) 481.

[23] Cecil F. Powell, "Fragments of autobiography," *Selected papers of Cecil Frank Powell*, eds. E.H.S. Burhop, W.O. Lock, M.G.K. Menon (Amsterdam: North Holland Publishing Company, 1972), 23.

[24] Werner Heisenberg, "Der Durchgang sehr energiereicher Korpuskeln durch den Atomkern," *Naturwissenschaften* 25 (1937) 749; idem, *Ber. Sächs. Akad. Wiss., Math.-Phys. Kl.* 89 (1937) 369.

[25] Hans Euler and Werner Heisenberg, "Theoretische Gesichtspunkte zur Deutung der kosmischen Strahlung," *Ergebn. Exakt. Naturwiss.* 17 (1938) 1.

[26] Werner Heisenberg, "Zur Theorie der 'Schauer' in der Höhenstrahlung," *Z. Physik* 101 (1936) 533.

[27] Karl von Meyenn, ed., *Wolfgang Pauli – Wissenschaftlicher Briefwechsel mit Bohr, Einstein, Heisenberg u. a.* (Berlin: Springer-Verlag, 1985), Vol. II, 495, Vol. III, 7, 9, 29, 31.

[28] H. Ziegert, "Die genaue Messung der von einem einzelnen α-Teilchen erzeugten Ionenmengen und der Nachweis neuer Aktivitäten," *Z. Physik* 46 (1928) 668.

[29] Josef Schintlmeister, "Zur Frage der Existenz noch unbekannter natürlicher α-Strahler," *MIR* 371; *Sitzungsber. Akad. Wiss. Wien, Math. Naturwiss. Kl. IIa* 144 (1935) 475; "Untersuchung über den Ursprung der α-Strahlen von 2 cm Reichweite," *MIR* 383; *Sitzungsber. Akad. Wiss. Wien, Math. Naturwiss. Kl. IIa* 145 (1936) 446.

[30] Marietta Blau, autobiographical sketch (transmitted by Leopold Halpern, who had received it from Marietta Blau's brother Otto Blau in 1977), Sondersammlung Zentralbibl. Phys.

[31] R.V. Gentry, "Radiohalos: Some unique lead isotope ratios and unknown alpha-radioactivity," *Science* 173 (1971) 727.

[32] Murray Gell-Mann, "The eightfold way: A theory of strong-interaction symmetry," *California Institute of Technology Synchrotron Laboratory Report* CTSL 20, 1961; "Symmetries of baryons and mesons," *Phys. Rev.* 125 (1962) 1067.

[33] Yuval Ne'eman, "Derivation of strong interactions from a gauge invariance," *Nucl. Phys.* 26 (1961) 222.

[34] Murray Gell-Mann, "A schematic model of baryons and mesons," *Phys. Lett.* 8 (1964) 214.

[35] George Zweig, "An SU(3) model for strong-interaction symmetry and its breaking II," *CERN Rept.* 8419/Th 412, 1964.

[36] Marian Danysz and Jerzy Pniewski, "Delayed disintegration of a heavy nuclear fragment I," *Phil. Mag.* (7th Ser.) 44 (1953) 348.

Bibliographies

Abbreviation:

MIR ... Mitteilungen des Instituts für Radiumforschung

Marietta Blau's Scientific Publications in Chronological Order

B1 M. Blau, "Über die Absorption divergenter γ-Strahlung," *MIR* 110 (1918); *Sitzungsber. Akad. Wiss. Wien, Math. Naturwiss. Kl. IIa* 127 (1918) 1253–1279.

B2 M. Blau and K. Altenburger, "Über einige Wirkungen von Strahlen II," *Z. Phys.* 12 (1922) 315–329.

B3 M. Blau and K. Altenburger, "Über eine Methode zur Bestimmung des Streukoeffizienten und des reinen Absorptionskoeffizienten von Röntgenstrahlen," *Z. Phys.* 25 (1924) 200–214.

B4 M. Blau, "Über die Zerfallskonstante von RaA," *MIR* 161 (1924); *Sitzungsber. Akad. Wiss. Wien, Math. Naturwiss. Kl. IIa* 133 (1924) 17–22.

B5 M. Blau, "Über die photographische Wirkung natürlicher H-Strahlen," *MIR* 179 (1925); *Sitzungsber. Akad. Wiss. Wien, Math. Naturwiss. Kl. IIa* 134 (1925) 427–436.

B6 M. Blau, "Die photographische Wirkung von H-Strahlen aus Paraffin und Aluminium," *Z. Phys.* 34 (1925) 285–295.

B7 M. Blau and E. Rona, "Ionisation durch H-Strahlen," *MIR* 190 (1926); *Sitzungsber. Akad. Wiss. Wien, Math. Naturwiss. Kl. IIa* 135 (1927) 573–585.

B8 M. Blau, "Über die photographische Wirkung von H-Strahlen II," *MIR* 208 (1927); *Sitzungsber. Akad. Wiss. Wien, Math. Naturwiss. Kl. IIa* 136 (1928) 469–480.

B9 M. Blau, "Über die photographische Wirkung von H-Strahlen aus Paraffin und Atomfragmenten," *Z. Phys.* 48 (1928) 751–764.

B10 M. Blau, "Über photographische Intensitätsmessungen von Poloniumpräparaten," *MIR* 220 (1928); *Sitzungsber. Akad. Wiss. Wien, Math. Naturwiss. Kl. IIa* 137 (1928) 259–268.

B11 M. Blau and E. Rona, "Weitere Beiträge zur Ionisation durch H-Partikeln," *MIR* 241 (1929); *Sitzungsber. Akad. Wiss. Wien, Math. Naturwiss. Kl. IIa* 138 (1929) 717–731.

B12 M. Blau and E. Rona, "Anwendung der Chamié'schen photographischen Methode zur Prüfung des chemischen Verhaltens von Polonium," *MIR* 257 (1930); *Sitzungsber. Akad. Wiss. Wien, Math. Naturwiss. Kl. IIa* 139 (1930) 275–279.

B13 M. Blau, "Quantitative Untersuchung der photographischen Wirkung von α- und H-Partikeln," *MIR* 259 (1930); *Sitzungsber. Akad. Wiss. Wien, Math. Naturwiss. Kl. IIa* 139 (1930) 327–347.

B14 M. Blau and E. Kara-Michailova, "Über die durchdringende Strahlung des Poloniums," *MIR* 283 (1931); *Sitzungsber. Akad. Wiss. Wien, Math. Naturwiss. Kl. IIa* 140 (1931) 615–622.

B15 M. Blau, "Über photographische Untersuchungen mit radioaktiven Strahlungen," *Zehn Jahre Forschung auf dem physikalisch-medizinischen Grenzgebiet*, ed. F. Dessauer (Leipzig: Georg Thieme Verlag, 1931), 390–398.

B16 M. Blau, "Über das Abklingen des latenten Bildes bei Exposition mit α-Partikeln," *MIR* 284 (1931); *Sitzungsber. Akad. Wiss. Wien, Math. Naturwiss. Kl. IIa* 140 (1931) 623–628.

B17 M. Blau and H. Wambacher, "Über das Verhalten einer kornlosen Emulsion gegenüber α-Partikeln," *MIR* 291 b (1932); *Sitzungsber. Akad. Wiss. Wien, Math. Naturwiss. Kl. IIb* 141 (1932) 467–474; also *Monatsh. Chem.* 61 (1932) 99–106.

B18 M. Blau and H. Wambacher, "Über Versuche, durch Neutronen ausgelöste Protonen photographisch nachzuweisen," *MIR* 296 a (1932); *Anz. Akad. Wiss. Wien* 69 (1932) 180–181.

B19 M. Blau and H. Wambacher, "Über Versuche, durch Neutronen ausgelöste Protonen photographisch nachzuweisen II," *MIR* 299 (1932); *Sitzungsber. Akad. Wiss. Wien, Math. Naturwiss. Kl. IIa* 141 (1932) 617–620.

B20 M. Blau, "Eine neue Fremdabsorption in Alkalihalogenid-kristallen," *Nachr. Ges. Wiss. Göttingen II* 51 (1933) 401–405.

B21 M. Blau and H. Wambacher, "Über den Einfluß des Kornzustands auf die Schwärzungsempfindlichkeit bei Exposition mit α-Partikeln," *Z. Wiss. Photogr. Photophys. Photochem.* 31 (1933) 243–250.

B22 M. Blau, "La méthode photographique et les problèmes de désintégration artificielle des atomes," *J. Phys. Radium* (Serie 7) 5 (1934) 61–66.

B23 M. Blau and H. Wambacher, "Physikalische und chemische Untersuchungen zur Methode des photographischen Nachweises von H-Strahlen," *MIR* 339 (1934); *Sitzungsber. Akad. Wiss. Wien, Math. Naturwiss. Kl. IIa* 143 (1934) 285–301.

B24 M. Blau and H. Wambacher, "Versuche nach der photographischen Methode über die Zertrümmerung des Aluminiumkernes," *MIR* 344 (1934); *Sitzungsber. Akad. Wiss. Wien, Math. Naturwiss. Kl. IIa* 143 (1934) 401–410.

B25 M. Blau and H. Wambacher, "Zum Mechanismus der Desensibilisierung photographischer Platten," *Z. Wiss. Photogr. Photophys. Photochem.* 33 (1934) 191–197.

B26 M. Blau and H. Wambacher, "Die photographische Methode in der Atomforschung," *Photogr. Korresp.* 70, Suppl. 5 (1934) 31–40.

B27 M. Blau and H. Wambacher, "Photographic desensitisers and oxygen," *Nature* (London) 134 (1934) 538.

B28 M. Blau and H. Wambacher, "Über die Empfindlichkeit desensibilisierter photographischer Schichten in Abhängigkeit vom Luftsauerstoff und von der Konzentration der Desensibilisatoren," *MIR* 367 (1935); *Sitzungsber. Akad. Wiss. Wien, Math. Naturwiss. Kl. IIa* 144 (1935) 403–408.

B29 M. Blau and H. Wambacher, "Zum Mechanismus der Desensibilisierung photographischer Platten II," *Z. Wiss. Photogr. Photophys. Photochem.* 34 (1935) 253–266.

B30 M. Blau, "Über den Einfluß des Luftsauerstoffes auf den photographischen Prozeß der Ausbleichung," *Photogr. Korresp.* 71, Suppl. 3 (1935) 21–28.

B31 M. Blau and H. Wambacher, "Zur Frage der Verteilung der α-Bahnen der Radiumzerfallsreihe," *MIR* 387 (1936); *Sitzungsber. Akad. Wiss. Wien, Math. Naturwiss. Kl. IIa* 145 (1936) 605–609.

B32 M. Blau and H. Wambacher, "Über den desensibilisierenden Einfluss von Chlor- und Bromsalzlösungen auf mit Farbstoffen imprägnierte photographische Schichten," *Photogr. Korresp.* 72 (1936) 108–109.

B33 M. Blau and H. Wambacher, "Bemerkungen zur Desensibilisierungstheorie von K. Weber," *Z. Wiss. Photogr. Photophys. Photochem.* 35 (1936) 211–215.

B34 M. Blau and H. Wambacher, "Längenmessung von H-Strahlbahnen mit der photographischen Methode," *MIR* 397 (1937); *Sitzungsber. Akad. Wiss. Wien, Math. Naturwiss. Kl. IIa* 146 (1937) 259–272.

B35 M. Blau and H. Wambacher, "Vorläufiger Bericht über photographische Ultrastrahlenuntersuchungen nebst einigen Versuchen über die 'spontane Neutronenemission'. Auftreten von H-Strahlen ähnlichen Bahnen entsprechend mehreren Metern Reichweite in Luft," *MIR* 404 (1937); *Sitzungsber. Akad. Wiss. Wien, Math. Naturwiss. Kl. IIa* 146 (1937) 469–477.

B36 M. Blau and H. Wambacher, "Disintegration processes by cosmic rays with the simultaneous emission of several heavy particles," *Nature* (London) 140 (1937) 585.

B37 M. Blau (after joint experiments with H. Wambacher), "Über die schweren Teilchen in der Ultrastrahlung" (abstract of lecture), *Verh. der Deutschen Phys. Ges.* 18 (1937) 123.

B38 M. Blau and H. Wambacher, "II. Mitteilung über photographische Untersuchungen der schweren Teilchen in der kosmischen Strahlung. Einzelbahnen und Zertrümmerungssterne," *MIR* 409 (1937); *Sitzungsber. Akad. Wiss. Wien, Math. Naturwiss. Kl. IIa* 146 (1937) 623–641.

B39 M. Blau, "Photographic tracks from cosmic rays," *Nature* (London) 142 (1938) 613.

B40 M. Blau and H. Wambacher, "Die photographische Methode in der Atomforschung. II. Bericht," *Photogr. Korresp.* 74 (1938) 2–6 and 23–29.

B41 M. Blau, "Über das Vorkommen von Alpha-Teilchen mit Reichweiten zwischen 1.2 und 2 cm in einer Samariumlösung," *Arch. Math. Naturvidensk.* B 42 (4) (1939) 1–10.

B42 M. Blau, "Sobre la existencia de una radiación α cuyo origen hasta ahora se desconoce," *Ingeniería* (Mexico) 14 (1940) 50–56.

B43 M. Blau, "El helio. Su origen y su localización," *Ciencia* (Mexico) 1 (1940) 265–270.

B44 M. Blau, "La radiación solar en las condiciones de México," *Ciencia* (Mexico) 3 (1942) 149–157.

B45 M. Blau, "Investigation of the radioactivity of rocks and thermal springs in Mexico," *Yearbook Am. Phil. Soc.* 1943, 134–135.

B46 M. Blau, "Algunas investigaciones sobre radiactividad llevadas a cabo en México," *Ciencia* (Mexico) 5 (1944) 12–17.

B47 M. Blau, "Notas para la medición de pequeñas corrientes de ionización," *Rev. Mex. Electr.*, April 1944, 5–7.

B48 M. Blau, "La radiactividad y el estado térmico de la tierra," *Ciencia* (Mexico) 5 (1944) 97–103.

B49 M. Blau and B. Dreyfus, "The multiplier photo-tube in radioactive measurements," *Rev. Sci. Instrum.* 16 (1945) 245–248.

B50 M. Blau and I. Feuer, "Radioactive light sources," *J. Opt. Soc. Am.* 36 (1946) 576–580.

B51 M. Blau, H. Sinason, and O. Baudisch, "Radioactivation of colloidal gamma ferric oxide," *Science* 103 (1946) 744–748.

B52 M. Blau and J. Carlin, "Ionization currents from extended alpha-sources," *Rev. Sci. Instrum.* 18 (1947) 715–721.

B53 M. Blau and H. Sinason, "Routine analysis of the alpha activity of protactinium samples," *Science* 106 (1947) 400–401.

B54 M. Blau and J.R. Carlin, "Industrial applications of radioactivity," *Electronics* 21 (1948) 78–82.

B55 M. Blau and J.E. Smith, "Beta-ray measurements and units," *Nucleonics* 2 (6) (1948) 67–74.

B56 M. Blau and J.A. De Felice, "Development of thick emulsions by a two-bath method," *Phys. Rev.* 74 (1948) 1198.

B57 M. Blau, "Grain density in photographic tracks of heavy particles," *Phys. Rev.* 75 (1949) 279–282.

B58 M. Blau, "Grain density in nuclear tracks" (abstract of lecture), *Phys. Rev.* 75 (1949) 1327.

B59 M. Blau, M.M. Block, and J.E. Nafe, "Heavy particles in cosmic-ray stars," *Phys. Rev.* 76 (1949) 860–861.

B60 M. Blau, I.W. Ruderman, and J. Czechowski, "Photographic methods of measuring slow neutron intensities," *Rev. Sci. Instrum.* 21 (1950) 232–236.

B61 M. Blau, "Bericht über die Entdeckung der durch kosmische Strahlung erzeugten 'Sterne' in photographischen Emulsionen," *Sitzungsber. Österr. Akad. Wiss., Math. Naturwiss. Kl. IIa* 159 (1950) 53–57.

B62 R. Rudin, M. Blau, and S. Lindenbaum, "A semi-automatic device for analyzing events in nuclear emulsions" (abstract of lecture), *Phys. Rev.* 78 (1950) 319–320.

B63 M. Blau, J. Nafe, and H. Bramson, "The dependance of high altitude star and meson production rates on absorbers" (abstract of lecture), *Phys. Rev.* 78 (1950) 320.

B64 M. Blau, "Möglichkeiten und Grenzen der photographischen Methode in Kernphysik und kosmischer Strahlung," *Acta Phys. Austriaca* 3 (1950) 384–395.

B65 M. Blau, R. Rudin, and S. Lindenbaum, "Semi-automatic device for analyzing events in nuclear emulsions," *Rev. Sci. Instrum.* 21 (1950) 978–985.

B66 M. Blau and A.R. Oliver, "Stars induced by high energy neutrons in the light elements of the photographic emulsion" (abstract of lecture), *Phys. Rev.* 87 (1952) 182.

B67 M. Blau and E.O. Salant, "T-Tracks in nuclear emulsions," *Phys. Rev.* 88 (1952) 954–955.

B68 M. Blau, A.R. Oliver, and J.E. Smith, "Neutron and meson stars induced in the light elements of the emulsion," *Phys. Rev.* 91 (1953) 949–957.

B69 M. Blau, M. Caulton, and J.E. Smith, "Meson production by 500-MeV negative pions," *Phys. Rev.* 92 (1953) 516–517.

B70 M. Caulton, M. Blau, and J.E. Smith, "Interactions of 500-MeV negative pions with emulsion nuclei" (abstract of lecture), *Phys. Rev.* 93 (1954) 919.

B71 M. Blau and M. Caulton, "Inelastic scattering of 500-MeV negative pions in emulsion nuclei," *Phys. Rev.* 96 (1954) 150–160.

B72 M. Blau and A.R. Oliver, "Interaction of 750-MeV π^- mesons with emulsion nuclei," *Phys. Rev.* 102 (1956) 489–494.

B73 M. Blau, "Hyperfragments and slow K-mesons in stars produced by 3-BeV protons," *Phys. Rev.* 102 (1956) 495–501.

B74 M. Blau, "Ionisationsmessungen in photographischen Emulsionen," *Acta Phys. Austriaca* 12 (4) (1959) 336–355.

B75 M. Blau, C.F. Carter, and A. Perlmutter, "Negative pion interactions at 1.3 GeV/c," *Nuovo Cimento* 14 (1959) 704–721; also *PB Report* 145040, USDCTS (1959); also *Department of Defense Report* AFOSR-TN-59-788 (1959).

B76 M. Blau, S.C. Bloch, C.F. Carter, and A. Perlmutter, "Studies of ionization parameters in nuclear emulsions," *Rev. Sci. Instrum.* 31 (1960) 289–297; also *Department of Defense Report* AFOSR-TN-59-229 (1959).

B77 M. Blau, C.F. Carter, and A. Perlmutter, "Study of antiproton interactions," *Department of Defense Report* AFOSR-TN-60-461 (1960).

B78 M. Blau, C.F. Carter, and A. Perlmutter, "Interaction and decays of hyperons produced in K-capture stars at rest," *PB Report* 149322, USDCTS (1960); also *Department of Defense Report* AFOSR-TN-60-745 (1960).

B79 M. Blau, in *Methods of Experimental Physics – Vol 5: Nuclear Physics*, eds. L.C.L. Yuan and C.-S. Wu (New York and London: Academic Press, 1961, 1963).
  (1) Section 1.7. "Photographic emulsions," Vol. 5A, 208–264.
  (2) Section 2.1.1.3. "Charge determination of particles in photographic emulsions," Vol. 5A, 298–307.

(3) Section 2.2.1.1.5. "Momentum measurement in nuclear emulsions," Vol. 5A, 388–408.
(4) Section 2.2.3.8. "Detection and measurement of gamma-rays in photographic emulsions," Vol. 5A, 676–682.
(5) Section 2.3.5. "Determination of mass of nucleons in emulsions," Vol. 5B, 37–44.

B80 M. Blau, C.F. Carter, and A. Perlmutter, "An example of hyperfragment decay in the π^+ mode and other interactions of K$^-$ mesons and hyperons in emulsion," *Nuovo Cimento* 27 (1963) 774–785.

B81 M. Blau, "Fotografitseckie emulsii," *Principy i metody registracii elementarnich tschastits*, Moscow 1963 (partial translation of contributions in *Methods of Experimental Physics – Vol. 5: Nuclear Physics*).

SELECTED LITERATURE

Asper, Kathrin, *Verlassenheit und Selbstentfremdung*, dtv 15079 (Olten: Walter-Verlag AG, 1987).
Berger, C., *Teilchenphysik – eine Einführung* (Berlin, Heidelberg, New York: Springer-Verlag, 1992).
Berkley, George E., *Vienna and Its Jews. The Tragedy of Success 1880s–1980s* (Cambridge, Maryland: Abt Books; Lanham, Maryland: Madison Books, 1988).
Bernhard, Thomas, *Heldenplatz* (Frankfurt/Main: Suhrkamp Taschenbuch 2474, 1988).
Bischof, B., "Marietta Blau (1894–1970)," *Wissenschaft und Forschung in Österreich*, ed. Gerhard Heindl (Frankfurt/Main: Peter Lang – Europäischer Verlag der Wissenschaften, 2000), 147.
Botz, Gerhard, Oxaal, Ivar, and Pollak, Michael (eds.), *Eine zerstörte Kultur. Jüdisches Leben und Antisemitismus in Wien seit dem 19. Jahrhundert* (Buchloe: Druck und Verlag Obermayer GmbH, 1990).
Broda, Engelbert, *Ludwig Boltzmann, Mensch, Physiker, Philosoph*, 2nd ed. (Vienna: Franz Deuticke, 1986).
Broda, Engelbert, *Ludwig Boltzmann*, transls. Larry Gay and Engelbert Broda (Woodbridge, Conn.: Oxbow Press, 1983).

Chadwick, J., "Possible Existence of a Neutron," *Nature* 129 (1932) 312.
Chadwick, J., "The Existence of a Neutron," *Proc. Roy. Soc.* A 136 (1932) 692.
Chamié, C., "Sur les groupements d'atomes d'éléments radio-actifs dans le mercure," *C. R. Acad. Sci. Paris* 184 (1927) 1243; "Sur l'existence de groupements d'atomes de radio-éléments dans les solutions acides et sur les surfaces activées par l'émanation," *C. R. Acad. Sci. Paris* 185 (1927) 770, "Sur le phénomène de groupements d'atomes des radioéléments," *C. R. Acad. Sci. Paris* 185 (1927) 1277; "Sur le phénomène de groupements d'atomes pour les émanations et pour les mélanges des radioéléments," *C. R. Acad. Sci. Paris* 186 (1928) 1838.
Chamié, C., "Étude du phénomène des groupements d'atomes des radioéléments," *J. Phys. Radium* 10 (1929) 44.
Clare, George, *Last Waltz in Vienna* (London: Pan Macmillan Ltd., 2002).
Debye, P., and Hückel, E., "Zur Theorie der Elektrolyte," *Physikal. Z.* 24 (1923) 185.
Euler, H., and Heisenberg, W., "Theoretische Gesichtspunkte zur Deutung der kosmischen Strahlung," *Ergebn. exakt. Naturwiss.* 17 (1938) 1.
Fischer, K., *Changing Landscapes of Nuclear Physics: A Scientometric Study on the Social and Cognitive Position of German-speaking Emigrants within the Nuclear Physics Community, 1921–1947* (Berlin, New York: Springer, 1993).
Fischer, K., "Die Emigration deutschsprachiger Physiker 1933: Strukturen und Wirkungen," *Die Emigration der Wissenschaften nach 1933*, eds. Herbert A. Strauss et al. (Munich, etc.: Disziplingeschichtliche Studien, K. G. Saur, 1991), 25.
Galison, P.L., "Nuclear Emulsions – The Anxiety of the Experimenter," *Image and Logic – A Material Culture of Microphysics* (Chicago: Univ. Chicago Press, 1997), 143.
Galison, P.L., "Marietta Blau – Between Nazis and Nuclei," *Physics Today* 50 (1997) 42.

Gold, Hugo, *Gedenkbuch der untergegangenen jüdischen Gemeinden des Burgenlands* (Tel-Aviv: Edition Olamenu, 1970).
Griffiths, David, *Introduction to Elementary Particles* (New York: Wiley, 1987).
Grotelüschen, F., *Der Klang der Superstrings. Einführung in die Natur der Elementarteilchen* (München: Deutscher Taschenbuchverlag, 1999).
Haider, G., *Untersuchung der Wechselwirkungen von 10 GeV/c μ^--Mesonen in Emulsionen*, Dissertation, University of Vienna (1964).
Halpern, L., "Marietta Blau − Discoverer of the Cosmic Ray 'Stars'," *A Devotion to Their Science*, eds. M.F. Rayner-Canham and G.W. Rayner-Canham (Montreal & Kingston, London, Buffalo: McGill Queen's University Press, 1997), 196.
Halpern, L., "Marietta Blau (1894−1970)," *Women in Chemistry and Physics*, eds. L.S. Grinstein, R.K. Rose, and M.H. Rafailovich (Westport, Connecticut and London: Greenwood Press, 1993), 57.
Heisenberg, W., "Der Durchgang sehr energiereicher Korpuskeln durch den Atomkern," *Ber. Sächs. Akad. Wiss., Math.-Phys. Kl.* 89 (1937) 369.
Heisenberg, W., "Der Durchgang sehr energiereicher Korpuskeln durch den Atomkern," *Naturwissenschaften* 25 (1937) 749.
Hevesy, G. v., "Erinnerung an die alten Tage am Wiener Institut für Radiumforschung," *Sitzungsber. Österr. Akad. Wiss., Math. Naturwiss. Kl. IIa* 159 (1950) 47.
Hevesy, George, and Paneth, Fritz, *A Manual of Radioactivity* transl. Robert W. Lawson (London: Oxford University Press, 1926).
Jelley, Nicholas Alfred, *Fundamentals of Nuclear Physics* (Cambridge, New York, etc.: Cambridge University Press, 1990).
Karlik, B., "1938−1950," *Sitzungsber. Österr. Akad. Wiss., Math. Naturwiss. Kl. IIa* 159 (1950) 35.
Karlik, B., "Bericht aus dem Institut für Radiumforschung und Kernphysik," *Almanach Österr. Akad. Wiss.* 120 (1970) 200.

Karlik, B., and Schmid, E., *Franz S. Exner und sein Kreis* (Vienna: Verlag der Österreichischen Akademie der Wissenschaften, 1982).

Katscher, F., "Jüdische Naturwissenschaftler und Techniker Österreichs – Was sie der Welt schenkten," *Österreichisch-jüdisches Geistes- und Kulturleben*, ed. Liga der Freunde des Judentums (Vienna: Literas-Universitätsverlag, 1988), 80.

Kienzl, Heinz, *Ein Frühling wird in der Heimat blühen. Erinnerungen und Spurensuche* (Vienna: Deuticke, 2002).

Kirsch, G., "Über die Atomzertrümmerung durch α-Strahlen IV. Abbau von Stickstoff und Sauerstoff. – Helium als Abbauprodukt," *MIR* 169; *Sitzungsber. Akad. Wiss. Wien, Math. Naturwiss. Kl. IIa* 133 (1924) 461.

Kirsch, G., and Pettersson, H., "Experiments on the Artificial Disintegration of Atoms," *Phil. Mag.* (Ser. 6) 47 (1924) 500.

Kirsch, G., and Pettersson, H., "Über die Atomzertrümmerung durch α-Partikeln," *MIR* 160; *Sitzungsber. Akad. Wiss. Wien, Math. Naturwiss. Kl. IIa* 132 (1923) 299; "Über die Atomzertrümmerung durch α-Partikeln II. Eine Methode zur Beobachtung der Atomtrümmer von kurzer Reichweite," *MIR* 167; *Sitzungsber. Akad. Wiss. Wien, Math. Naturwiss. Kl. IIa* 133 (1924) 235.

Kloyber, Christian, and Patka, Marcus G., *Österreicher im Exil, Mexiko 1938–1945* (edited for Dokumentationsarchiv des österreichischen Widerstands) (Vienna: Deuticke, 2002).

Lauda, H., "Über das Abklingen des latenten Bildes auf der photographischen Platte," *MIR* 390; *Sitzungsber. Akad. Wiss. Wien, Math. Naturwiss. Kl. IIa* 145 (1936) 1.

Leng, H., "Zur Frage der photographischen Wirksamkeit sonnenbestrahlter Metalle," *MIR* 262a (1930).

Lichtenberger-Fenz, Brigitte, "Österreichs Hochschulen – Opfer oder Wegbereiter der nationalsozialistischen Gewaltherrschaft? (Am Beispiel der Universität Wien)," *Willfährige Wissenschaft: Die Universität Wien 1938–1945*, eds. Gernot Heiß et al. (Vienna: Verlag für Gesellschaftskritik, 1989).

Lohrmann, E., *Einführung in die Elementarteilchenphysik*, 2nd ed. (Stuttgart: B.G. Teubner, 1990).

Lucha, W., and Regler, M., *Elementarteilchenphysik in Theorie und Experiment* (Kufstein: Paul Sappl Verlag, 1997).
Marsden, E., "The Passage of α-Particles through Hydrogen," *Phil. Mag.* (Ser. 6) 27 (1914) 824.
Medawar, Jean, and Pyke, David, *Hitler's Gift. Scientists Who Fled Nazi Germany*. Foreword by Dr. Max Perutz (London: Piatkus, 2001).
Merhaut, Otto, *Das Problem der Resonanzeindringung von α-Teilchen in den Aluminiumkern, behandelt mit der photographischen Methode*, Dissertation, University of Vienna (1938).
Meyenn, K. v., ed., *Wolfgang Pauli – Wissenschaftlicher Briefwechsel mit Bohr, Einstein, Heisenberg u. a.* (Berlin: Springer-Verlag, 1985), Vol. II, 495; Vol. III, 7, 9, 29, 31.
Meyer, St., "Otto Hönigschmid," *Anzeiger Akad. Wiss. Wien* 82 (1945) 23.
Meyer, Stefan, "Vorgeschichte der Gründung und das erste Jahrzehnt des Institutes für Radiumforschung," *Sitzungsber. Österr. Akad. Wiss., Math. Naturwiss. Kl. IIa* 159 (1950) 1.
Meyer, St., and Schweidler, E., *Radioaktivität*, 2nd ed. (Leipzig and Berlin: Verlag B.G. Teubner, 1927).
Michl, W., "Über die Photographie der Bahnen einzelner α-Teilchen," *Sitzungsber. Kaiserl. Akad. Wiss. Wien, Math. Naturwiss. Kl. IIa* 121 (1912) 1431.
Michl, W., "Zur photographischen Wirkung der α-Teilchen," *Sitzungsber. Kaiserl. Akad. Wiss. Wien, Math. Naturwiss. Kl. IIa* 123 (1914) 1955.
Mitteilungen des Verbands der Akademikerinnen Österreichs 66, Special issue 3A (1997).
Moore, W., *Schrödinger – Life and Thought* (Cambridge: Cambridge University Press, 1989), 479.
Musiol, G., Ranft, J., Reif, R., and Seeliger, D., *Kern- und Elementarteilchenphysik* (Weinheim: Verlag Chemie, 1998).
Okun, Lev Borisovich, *α, β, γ ... Z: A Primer in Particle Physics* (Chur, London, Paris, New York: Harwood Academic Publishers, 1987).
Paneth, Fritz, *Radio-Elements as Indicators* (New York and London: McGraw-Hill, 1928).

Paneth, F.A., "Aus der Frühzeit des Wiener Radiuminstituts. Die Darstellung des Wismutwasserstoffs," *Sitzungsber. Österr. Akad. Wiss., Math. Naturwiss. Kl. IIa* 159 (1950) 49.

Patka, Marcus G., *Zu nahe der Sonne, Deutsche Schriftsteller im Exil in Mexiko* (Berlin: Aufbau Taschenbuchverlag, 1999).

Pauley, Bruce F., *From Prejudice to Persecution* (Chapel Hill: University of North Carolina Press, 1992). In German: *Eine Geschichte des österreichischen Antisemitismus*, transl. Helga Zoglmann (Vienna: Kremayr & Scheriau, 1993).

Perkins, Donald H., *Introduction to High Energy Physics*, 4th ed. (Cambridge: Cambridge University Press, 2000).

Pettersson, H., and Kirsch, G., *Atomzertrümmerung, Verwandlung der Elemente durch Bestrahlung mit α-Teilchen* (Leipzig: Akademische Verlagsgesellschaft, 1926).

Powell, C.F., Fowler, P.H., and Perkins, D.H., *The Study of Elementary Particles by the Photographic Method* (London: Pergamon Press, 1959), 35, 43–49, 50–52.

Powell, C.F., "Fragments of Autobiography," *Selected Papers of Cecil Frank Powell*, eds. E.H.S. Burhop, W.O. Lock, M.G.K. Menon (Amsterdam: North Holland Publishing Co., 1972), 23.

Przibram, K. in "Das 50jährige Bestandsjubiläum des Institutes für Radiumforschung," *MIR* 550; *Sitzungsber. Österr. Akad. Wiss., Math. Naturwiss. Kl. II* 170 (1960) 239.

Przibram, K., "1920–1938," *Sitzungsber. Österr. Akad. Wiss. IIa* 159 (1950) 27.

Przibram, K., "Stefan Meyer (Obituary)," *Almanach Österr. Akad. Wiss.* 100 (1950) 340.

Radvanyi, Pierre, *Jenseits des Stroms. Erinnerungen an meine Mutter Anna Seghers* (Berlin: Aufbau Verlag, 2005).

Reiter, Wolfgang L., "Naturwissenschaften und Remigration: Vertreibung ohne Rückkehr," *Austriaca, Cahiers universitaires d'information sur l'Autriche*, Université de Rouen, 2003.

Reiter, Wolfgang L., "Stefan Meyer: Pioneer of Radioactivity," *Physics in Perspective* 3 (2001) 106.

"Review of Particle Physics," *Phys. Rev.* D66, No. 1, Part I (2002).

Riedl, J., "Über die Gruppenstruktur der Rückstoßprotonen von α-Teilchen," *MIR* 416; *Sitzungsber. Akad. Wiss. Wien, Math. Naturwiss. Kl. IIa* 147 (1938) 181.

Romains, J., "Laudatio for M. Blau on the Occasion of Her Departure," *Austria Libre*, May 1944.

Rosner, R., "Der Ignaz-Lieben-Preis," *Chemie* 4 (1997) 30. For renewal of the prize, see the protocol of the plenary session of the Austrian Academy of Sciences, Vienna, April 2, 2004.

Rotblat, J., "Photographic Emulsion Technique," *Progr. Nucl. Phys.* 1 (1950) 37.

Rozenblit, M.L., *Die Juden Wiens 1867–1914. Assimilation und Identität* (Vienna: Böhlau, 1989).

Schorske, Carl E., *Fin de Siècle Vienna: Politics and Culture* (New York: Vintage Press, 1981).

Shapiro, M.M., "Nuclear Emulsions," *Handbuch der Physik*, Vol. 45, ed. S. Flügge (Berlin: Springer-Verlag, 1958), 342–436.

Shapiro, M.M., "Tracks of Nuclear Particles in Photographic Emulsions," *Rev. Mod. Physics* 13 (1941) 58.

Soukup, Werner R. (ed.), *Die wissenschaftliche Welt von gestern. Die Preisträger des Ignaz-L.-Lieben-Preises 1865–1937 und des Richard-Lieben-Preises 1912–1928* (Vienna: Böhlau, 2004).

Spitzer, Shlomo, *Die jüdische Gemeinde von Deutschkreutz* (Vienna: Böhlau, 1995).

Stadler, Friedrich (ed.), *Kontinuität und Bruch 1938 – 1945 – 1955, Beiträge zur österreichischen Kultur- und Wissenschaftsgeschichte* (Münster: Lit Verlag, 2004).

Stadler, Friedrich (ed.), *Vertriebene Vernunft* (Vienna and Munich: Verlag für Jugend und Volk, 1987).

Steinmaurer, R., "Erinnerungen an V.F. Hess, den Entdecker der kosmischen Strahlung, und an die ersten Jahre des Betriebes des Hafelekar-Labors," *Early History of Cosmic Ray Studies*, eds. Y. Sekido and H. Elliot (Dordrecht, Boston, London: D. Reidel, 1985), 28.

Steppan, E., "Das Problem der Zertrümmerung von Aluminium behandelt mit der photographischen Methode," *MIR* 370; *Sitzungsber. Akad. Wiss. Wien, Math. Naturwiss. Kl. IIa* 144 (1935) 455.

Stetter, G., in "Personal- und Hochschulnachrichten," *Österr. Chemiker-Zeitung* 51 (1950) 234.
Stetter, G., and Wambacher, H., "Neuere Ergebnisse von Untersuchungen über die Mehrfachzertrümmerung von Atomkernen durch Höhenstrahlen," *Physikal. Z.* 40 (1939) 702.
Stetter, G., and Wambacher, H., "Versuche zur Absorption der Höhenstrahlung nach der photographischen Methode I: Zertrümmerungssterne unter Blei-Absorption," *Sitzungsber. Akad. Wiss. Wien, Math. Naturwiss. Kl. IIa* 152 (1944) 1.
Stiftungsbrief Institut für Radiumforschung, *Almanach Kaiserl. Akad. Wiss.* 61 (1911) 212.
Stuewer, R.H., "Artificial Disintegration and the Cambridge-Vienna Controversy," *Observation, Experiment and Hypothesis in Modern Physical Science*, eds. Peter Achinstein and Owen Hannaway (Cambridge, Mass., London: MIT Press, 1985), 239.
Taylor, H.J., "The Tracks of α-Particles and Protons in Photographic Emulsions," *Proc. Roy. Soc.* A 150 (1935) 382.
Verduzco Ríos, E., "La ESIME, Un refugio en México para Marietta Blau," *Investigación hoy* 95 (2000) 50.
Wambacher, H., "Mehrfach-Zertrümmerung von Atomkernen durch kosmische Strahlung," *Angew. Chemie* 52 (1939) 117.
Wambacher, H., "Kernzertrümmerung durch Höhenstrahlung in der photographischen Emulsion," *MIR* 435; *Sitzungsber. Akad. Wiss. Wien, Math. Naturwiss. Kl. IIa* 149 (1940) 157.
Wambacher, H., "Mehrfachzertrümmerung von Atomkernen durch kosmische Strahlung; Ergebnisse aus 154 Zertrümmerungssternen in photographischen Platten," *Phys. Z.* 39 (1938) 883; *Z. techn. Physik* 19 (1938) 569.
Wambacher, H., "Über ein sicher identifiziertes Teilchen aus einer Höhenstrahlzertrümmerung," *Sitzungsber. Akad. Wiss. Wien, Math. Naturwiss. Kl. IIa* 154 (1945) 66.
Wambacher, H., "Untersuchung der photographischen Wirkung radioaktiver Strahlungen auf mit Chromsäure und Pinakryptolgelb vorbehandelte Filme und Platten," *MIR* 274; *Sitzungsber. Akad. Wiss. Wien, Math. Naturwiss. Kl. IIa* 140 (1931) 271.

Wambacher, H., and Widhalm, A., "Über die kurzen Bahnspuren in photographischen Schichten," *Sitzungsber. Akad. Wiss. Wien, Math. Naturwiss. Kl. IIa* 152 (1944) 173.

Widhalm, A., "Schwere Teilchen in der kosmischen Höhenstrahlung," *Z. Physik* 115 (1940) 481.

Zila, St., "Beiträge zum Ausbau der photographischen Methode für Untersuchungen mit Protonenstrahlen," *MIR* 386; *Sitzungsber. Akad. Wiss. Wien, Math. Naturwiss. Kl. IIa* 145 (1936) 503.

Zimmel, Brigitte, and Kerber, Gabriele, *Hans Thirring. Ein Leben für Physik und Frieden* (Vienna: Böhlau, 1992).

Zweig, Stefan, *The World of Yesterday*, transls. presumably Eden Paul and Cedar Paul (Lincoln, Nebraska: University of Nebraska Press, 1964).

INTERNET SOURCES

Byers, N., "Contributions of Twentieth Century Women to Physics," http://www.physics.ucla.edu/~cwp/phase2/Blau,_Marietta@843727247.html.

Web Site of The Nobel Foundation, http://www.nobel.se/laureate.

INDEX OF NAMES

Achinstein, Peter 114, 203
Altenburger, Kamillo 21, 151, 189
Alvin, Carl 64
Anderson, Carl D. 107, 172
Antropoff, Andreas von 38, 116
Aschner, Joseph 6, 113
Aschner, Helene, née Pallester 96, 113
Asper, Kathrin 196

Bader, Alfred 44
Bagge, Erich 48, 117, 162
Bartels, Hans 69
Batiz, Juan de Dios 53
Baudisch, Oskar 193
Becquerel, Henri 18, 145
Berger, C. 196
Berkley, George E. 196
Bernardini, Gilberto 85, 123, 144
Bernert, Traude 105
Bernhard, Thomas 196
Bethe, Hans 48, 117, 162
Bischof, Brigitte 112, 196
Bitter, Francis 133
Bitter, John 133
Bizberg, Paula 6, 59
Blackett, Patrick 150, 186
Blau, Eva, see Connors, Eva, née Blau
Blau, Florentine, née Goldenzweig 14, 15, 20, 58, 59, 60, 67, 70, 112
Blau, Fritz 15

Blau, Julius (Judah) 15
Blau, Lily 96
Blau, Ludwig 15, 17, 20, 55, 70, 72, 96, 97, 101, 112
Blau, Markus (Mayer) 13, 14, 15, 20, 113
Blau, Otto 15, 17, 20, 55, 70, 79, 87, 91, 100, 101, 112, 115, 124, 187
Bloch, Sylvan C. 7, 8, 90, 136, 137, 144, 195
Block, Martin M. 7, 8, 74, 130, 144, 194
Blomberg, Werner von 50
Bohr, Niels 51, 187, 200
Boltzmann, Ludwig 103, 104, 196
Booth, Eugene T. 130, 144
Bormann, Elisabeth 113
Born, Max 37, 124
Bosquez, Gilberto 63
Bothe, Walther 155
Botz, Gerhard 196
Bragg, William 31
Bramson, H. 194
Breunlich, Wolfgang 6, 125
Broda, Engelbert 196
Broglie, Louis de 31, 41
Bruno, Karl 16
Bucky, Gustav 49, 118
Burhop, Eric H.S. 187, 201
Buschbeck, Brigitte, née Czapp 7, 8, 94, 139, 140
Byers, Nina 112, 204

Cárdenas, Lázaro 54
Carlin, J.R. 193
Carter, Claude F. 132, 135, 137, 195, 196
Chadwick, James 27, 29, 35, 42, 150, 155, 185, 186, 197
Chalupka, Alfred 6, 7, 124
Chamié, Catherine 190, 197
Clare, George 197
Connors, Eva, née Blau 7, 112, 113, 119, 121, 123, 124
Coulomb, Charles A. de 158
Curie, Irène, see Joliot, Irène, née Curie
Curie, Marie 10, 23, 40, 41, 42, 100, 107, 116, 146, 156
Czapek, Gerhard 125
Czapp, Brigitte, see Buschbeck, Brigitte, née Czapp
Czechowski, J. 194

Danysz, Marian 188
Debye, Petrus 66, 197
De Felice, J.A. 194
Deutsch, Leo 64
Deutsch de Lechuga, Ruth 65
Dick, Auguste 114
Dirac, Maurice 118
Djerassi, Carl 14
Dollfuss, Engelbert 24, 43, 114
Dresel, Elisabeth 119
Dreyfus, B. 193
Duby-Blom, Gertrude 65
Dürer, Albrecht 142
Dvorak, Paul F. 7

Eggert, John 36, 39, 40, 115
Eggstain, Hannelore, see Sexl, Hannelore, née Eggstain

Ehrenhaft, Felix 17, 68, 83, 120
Einstein, Albert 5, 48, 49, 50, 53, 54, 60, 61, 64, 66, 118, 119, 163, 187, 200
Elliot, Harry 117, 202
Ellis, Hanne, née Lauda 7, 8, 96, 117, 127, 144, 154, 186, 199
Escherich, Gustav von 17
Esterházy, Fürst Paul 14
Eucken, Arnold 38, 116
Euler, Hans 187, 197
Exner, Franz S. 18, 103, 104, 106, 107, 123, 126, 199

Fabry, Charles 42
Feierl, Hedy 7
Felber, Heinz 125
Fermi, Enrico 72, 121
Feuer, Irving 193
Fierz, Markus 48, 162
Firnberg, Hertha 13
Fischer, Klaus 118, 197
Flamm, Ludwig 125
Fowler, P.H. 201
Føyn, E. 51
Franck, James 37, 38, 115
Franco, Francisco 63
Frei, Bruno 64
Freud, Sigmund 14
Frisch, Otto R. 51, 138
Frischauf-Pappenheim, Marie 64
Fritsch, Werner von 50
Furtwängler, Philipp 17

Gable, Barbara 7
Galison, Peter L. 112, 197
Gay, Larry 196

Gell-Mann, Murray 182, 188
Gentry, R.V. 163, 187
Gleditsch, Ellen 48, 49, 50, 51, 55, 56, 107, 108, 117, 118
Gold, Hugo 198
Goldhaber, Maurice 84
Golwig, Hans 92
Golwig, Hugo 55
Griffiths, David 198
Grinstein, Luise S. 112, 198
Grön, Ortrud 125
Gross, Philipp 57
Grotelüschen, F. 198

Hahn, Otto 120
Haider, Gerda, see Petkov, Gerda, née Haider
Hainisch, Marianne 31, 114
Hainisch, Michael 114
Halpern, Leopold 7, 8, 55, 98, 99, 112, 115, 118, 124, 125, 137, 144, 187, 198
Hanffstengel, Renate 6
Hannaway, Owen 114, 203
Hanslik, Rudolf 126
Haschek, Edward 82, 123
Hasenöhrl, Friedrich 108
Havas, Peter 75, 122
Haworth, Leland John 84
Heindl, Gerhard 112, 196
Heinrich, Margarethe 6, 123
Heisenberg, Werner 48, 66, 140, 155, 162, 187, 197, 198, 200
Heiß, Gernot 199
Heitler, Walter 37, 39, 77, 115, 161
Herzog, Richard 120
Hess, Viktor 21, 44, 45, 52, 87, 107, 117, 157, 158, 159, 202

Hevesy, Georg von 21, 198
Hille, Peter 125
Hitler, Adolf 41, 49, 50, 200
Hlawka, Edmund 6
Holzknecht, Guido 20
Hönigschmid, Otto 21, 33, 101, 115, 200
Houtermans, Friedrich G. 142
Hückel, E. 197

Jacobi, Maria 124
Jäger, Gustav 103
Jelley, Nicholas A. 198
Johns, Jorun 7, 113
Joliot, Frédéric 40, 41, 155
Joliot, Irène, née Curie 40, 41, 155
Jong, Erica 11

Kara-Michailova, Elisabeth 190
Karlik, Berta 23, 26, 27, 29, 30, 31, 36, 37, 39, 41, 45, 50, 51, 52, 54, 55, 81, 82, 85, 86, 92, 96, 97, 98, 100, 101, 102, 105, 114, 116, 118, 123, 124, 126, 127, 128, 141, 142, 143, 168 198, 199
Katscher, Friedrich 112, 199
Kellner, Gottfried 125
Kerber, Gabriele 204
Kerber, Wolfgang 6, 122
Kienzl, Heinz 199
Kinoshita, S. 147, 185
Kirsch, Gerhard 26, 28, 29, 39, 52, 69, 80, 106, 150, 186, 199, 201
Kisch, Egon Erwin 64, 119
Kisch, Gisl 64
Kloyber, Christian 119, 199

Knoll, Fritz 125
Koczy, Fritz 90, 133
Kohn, Gustav 17
Kottler, Friedrich 68
Kupelwieser, Karl 18
Kursunoglu, Behram 134
Kurz, Trude 64
Kusch, Polykarp 72, 121, 122
Kutschera, Walter 6, 8, 11

Ladenburg, Rudolf 50, 118
Lamb, Willis 72, 122
Lattes, Cesare M.G. 135
Lauda, Hanne, see Ellis, Hanne, née Lauda
Lawson, Robert W. 198
Lecher, Erich 17
Lee, Tsung-Dao 111
Leng, Herta 32, 35, 75, 114, 115, 137, 199
Lewis, Sinclair 71
Lichtenberger-Fenz, Brigitte 199
Lieben, Ignaz L. 44
Lindenbaum, Seymour J. 7, 8, 75, 130, 144, 176, 194
Lindh, Axel E. 78, 123
Lock, W.O. 187, 201
Loewe, Franziska 7, 118
Lohrmann, E. 199
Lorentz, Hendrik A. 72, 121
Lucha, Wolfgang 200
Luft, Fritz 36

Machatschki, Felix 125
Mahler, Gustav 20
Marsden, Ernest 148, 185, 200
Matouschek, Franz 16

Mattauch, Josef 69, 120
Medawar, Jean 200
Meder, Ingrid 6
Meister, Richard 125
Meitner, Lise 10, 11, 14, 44, 69, 82, 138, 141
Menon, M.G.K. 187, 201
Merhaut, Otto 117, 156, 186, 200
Meschkan, Margarete 6, 16, 113
Meyenn, Karl von 187, 200
Meyer, Agathe 42
Meyer, Stefan 17, 18, 21, 23, 24, 29, 32, 36, 37, 38, 40, 41, 42, 44, 45, 47, 51, 52, 55, 67, 68, 69, 70, 71, 74, 81, 82, 103, 113, 115, 116, 117, 119, 120, 121, 122, 123, 141, 142, 200, 201
Michl, Wilhelm K. 28, 114, 147, 148, 152, 185, 200
Miller, Henry 9, 11
Millikan, Robert A. 120
Moore, Walter 124, 200
Morrison, Douglas R.O. 141
Musiol, G. 200
Mussolini, Benito 43

Nafe, J.E. 194
Ne'eman, Yuval 188
Nowotny, Hans 125

Oberkofler, Gerhard 6
Occhialini, Giuseppe P.S. 122, 135
Okun, Lev B. 200
Oliver, Anne R. 194, 195

Ortner, Gustav 68, 69, 80, 81, 82, 83, 97, 109, 110, 120, 123
Oxaal, Ivar 196

Paalen, Wolfgang 64
Paneth, Friedrich A. 44, 46, 47, 51, 52, 54, 55, 56, 117, 119, 198, 200, 201
Patka, Marcus G. 199, 201
Paul, Cedar 204
Paul, Eden 204
Paul, Wolfgang 83
Pauley, Bruce F. 201
Pauli, Wolfgang 48, 117, 162, 187, 200
Pearlstein, Sol 124
Perkins, Donald H. 135, 201
Perlmutter, Arnold 6, 7, 8, 89, 90, 98, 99, 110, 119, 122, 124, 125, 132, 137, 144, 181, 183, 195, 196
Perlmutter, Bernard 133
Perlmutter, Joseph 133
Perlmutter, Ruth 133
Perutz, Max 200
Peter, Gustav 49, 53
Petkov, Gerda, née Haider 7, 8, 94, 95, 99, 119, 125, 139, 141, 143, 144, 198
Pettersson, Hans 26, 27, 28, 29, 30, 39, 50, 51, 106, 114, 150, 186, 199, 201
Pfaundler, Leopold 104, 126
Pfleger, Christine, née Lohse 96
Pietschmann, Herbert 6, 7, 8, 95, 111, 140, 144
Planck, Max 66, 72, 121
Pniewski, Jerzy 188

Pohl, Robert W. 36, 37, 38, 39, 41, 42, 115, 142
Pollak, Michael 196
Pollak-Rudin, Robert 75, 131, 176, 194
Powell, Cecil F. 5, 10, 76, 77, 78, 86, 87, 110, 122, 134, 135, 161, 187, 201
Pregel, Boris 121
Przibram, Karl 23, 24, 41, 44, 51, 67, 68, 82, 93, 96, 104, 113, 114, 116, 120, 125, 126, 184, 201
Pyke, David 200

Rabi, Isidor I. 72, 121
Rabinowitsch, Eugen 38, 116
Radvanyi, Pierre 7, 8, 64, 65, 119, 128, 144, 201
Rafailovich, Miriam H. 112, 198
Rainwater, James 130
Ramsey, William 106
Ranft, J. 200
Rayner-Canham, Geoffrey W. 112, 198
Rayner-Canham, Marylene F. 112, 198
Regener, Erich 46
Regler, Friedrich M. 125
Regler, Meinhard 200
Reif, R. 200
Reinerová, Lenka 64
Reinganum, M. 147, 185
Reinhardt, Max 64
Reiter, Wolfgang L. 6, 112, 113, 114, 201
Richter, Elise 30, 114
Rieder, Fritz 39

Riedl, Johanna 117, 202
Ringelnatz, Joachim 97
Rivera, Diego 63, 64
Robitschek, Ernst 64
Rodhe, Agnes 7, 29, 114
Romains, Jules 67, 120, 202
Romatnik, Helmut 125
Römer, Ernst 63, 64
Römer, Irma 64
Rona, Elisabeth 29, 31, 33, 42, 50, 56, 75, 105, 114, 118, 127, 133, 189, 190
Röntgen, Wilhelm C. 145
Rose, Rose K. 112, 198
Rosenblum, Salomon 41, 116
Rotblat, Joseph 202
Rozenblit, Marsha L. 113, 202
Rozental, Stefan 138
Rubin, Marcel 64
Ruderman, Irvin W. 194
Rudin, Robert, see Pollak-Rudin, Robert
Rutherford, Ernest 26, 27, 29, 31, 110, 148, 150, 185

Salant, Edward O. 84, 85, 194
Sandoval Vallarta, Manuel 66, 129
Saulich, Anna 125
Schintlmeister, Josef 52, 162, 187
Schirmann, Maria Anna 113
Schlemko, Anny, née Frantz 6, 123
Schlögl, Reinhard 7, 124, 125, 144
Schmid, Erich 106, 125, 126, 199
Schnitzler, Arthur 14

Schönfeld, Thomas 6, 8
Schorske, Carl E. 202
Schrödinger, Erwin 66, 76, 77, 78, 87, 91, 118, 122, 124, 200
Schulmann, Robert 6
Schwarz, Theo 119
Schweidler, Egon von 18, 82, 101, 103, 115, 200
Schwerer, Otto 6
Seeliger, D. 200
Seghers, Anna 64, 65, 76, 201
Seidl, Franziska 52
Sekido, Yataro 117, 202
Sexl, Hannelore, née Eggstain 7, 8, 27, 34, 88, 95, 96, 97, 124, 125, 141, 143, 144
Sexl, Roman 27, 34, 88, 96, 97
Shapiro, M.M. 202
Sinason, H. 193
Šlibar, Neva 7
Smekal, Adolf G. 38, 116
Smith, J.E. 194, 195
Soukup, Werner R. 202
Spira, Steffi 64
Spitzer, Shlomo 202
Stadler, Friedrich 112, 114, 119, 202
Steinke, Eduard G. 37, 39, 45, 78, 115
Steinmaurer, Rudolf 45, 117, 202
Steppan, Elvira 116, 156, 186, 202
Stetter, Georg 45, 52, 68, 69, 80, 81, 82, 83, 97, 109, 120, 121, 123, 161, 187, 203
Strauss, Herbert A. 118, 197
Stuewer, Roger H. 114, 203

Taylor, H.J. 203
Thirring, Hans 37, 52, 68, 75, 81, 82, 87, 108, 122, 123, 124, 204
Thirring, Walter 6, 78, 94, 95, 122, 138, 139
Thomson, Joseph J. 104
Tinus, Auguste 16
Toeman, Richard 6, 119
Trotsky, Leon 63
Tyndall, A.M. 110

Verduzco Ríos, Esperanza 7, 118, 119, 203
Volk, Else 64
Volk, Richard 64
Vonach, Waltraut, née Patzak 125

Wagner, Franziska 94
Wagner, Georg 81, 123
Wambacher, Hertha 5, 10, 33, 34, 35, 36, 39, 40, 44, 45, 48, 55, 69, 76, 77, 78, 79, 80, 81, 83, 94, 101, 115, 120, 121, 123, 127, 128, 142, 154, 155, 156, 157, 159, 160, 161, 162, 183, 186, 187, 190, 191, 192, 193, 203, 204

Weinberger, Franz 17
Weinberger, Josef 20, 79
Weinberger, Katharina 59
Weinberger, Margarethe 60
Weisskopf, Viktor F. 138
Wessely, Friedrich 125
Widhalm, Anton 121, 161, 187, 204
Wilson, Charles T.R. 110
Winzeler, Helmut 141
Wu, Chien-Shiung 92, 111, 135, 184, 195

Yang, Chen Ning 111
Yuan, Luke C.L. 92, 135, 184, 195
Yukawa, Hideki 172

Zeuch, William E. 118
Ziegert, H. 187
Zila, Stefanie 116, 204
Zimmel, Brigitte 204
Zweig, George 188
Zweig, Stefan 204
Zwins, Hugo 125

Index of Subjects

Academy of Sciences
  Austrian (Imperial, Viennese, ÖAW) 7, 11, 18, 19, 22, 24, 26, 30, 31, 44, 47, 68, 69, 93, 102, 103, 104, 105, 109, 112, 117, 124, 125, 126, 139, 141, 199, 202, 223
  German 91
  Berlin-Brandenburg 7, 74, 124
  Royal Swedish 7, 78, 122, 123
Accelerator (see also Bevatron, Cockcroft generator, Cosmotron, Cyclotron) 73, 75, 85, 162, 173, 175, 176, 178
Acción Republicana Austriaca de México (ARAM) 63, 64, 67, 76
Acta Physica Austriaca 102, 123, 135, 194, 195
Agfa 35, 36, 38, 115
Airship (Zeppelin) 55
Albert Einstein Archive 7, 118, 119
Alpha-particle (α-particle) 10, 27, 28, 33, 35, 40, 146, 147, 148, 149, 151, 152, 153, 154, 155, 156, 157, 161, 162, 163, 165, 171, 174, 178, 179, 185, 190, 200, 203
Alpha-radiation (α-radiation) 32, 44, 52, 147, 154, 162, 163, 193

Alpha-rays (α-rays) 26, 27, 31, 34, 103, 146
Aluminum 32, 116, 117, 152, 154, 170, 186, 189, 191, 200, 202
American Association of University Women 31, 53
Antiproton 90, 135, 182, 195
Anti-Semitism 20, 24, 196, 201
Archive Radiumforschung, Austrian Academy of Sciences 86, 113, 115, 116, 117, 118, 119, 121, 122, 123, 124
Argonne National Laboratory 136
Association for the Extended Education of Women 15, 16
Atom Institute of the Austrian Universities 97, 110
Atomic disintegration (see also Nuclear fission) 26, 50, 76, 149, 199
Atomic Energy Commission 73, 75, 173, 174
Austrian Association of University Women (Verband der Akademikerinnen Österreichs, VAÖ) 30, 31, 36, 105, 114, 200
Austrian Treaty of State 13, 84
Austro-Hungarian Empire 18, 20
Autonomous University of Mexico (UNAM) 65, 66

Bad Ischl 21, 68, 103, 104, 123
Baden-Baden 69, 79

Index of Subjects 213

Balloon flight 46, 47, 52, 55, 107, 158
Baryon 175, 176, 187, 188
Berkeley 111, 135, 173, 182
Berlin 7, 13, 21, 29, 46, 66, 68, 69, 91, 128, 196, 197, 200, 201, 202
Bern 94, 95, 122, 139, 141, 142, 143
Beryllium 35, 40, 155, 156, 162, 165
Beta-particle (β-particle) 146
Beta-radiation (β-radiation) 34, 146, 152, 154, 171
Beta-rays (β-rays) 26, 172
Bevatron 73, 135, 173, 182
Blob density 182
Böhlau-Verlag 5, 113, 202, 204
Bohr Institute 50, 51, 118
Boltzmanngasse 18, 50, 94, 142
Breslau 16
Bristol 76, 79, 86, 87, 110, 115, 124, 161
Brookhaven National Laboratory (BNL) 73, 75, 76, 84, 85, 88, 99, 130, 136, 137, 173, 174, 178, 179, 180, 181
Bubble chamber 131, 134, 173, 181, 185

Cambridge 29, 31, 42, 87, 104, 110, 114, 122, 124, 148, 150, 155, 196, 198, 200, 201, 203
Canadian Radium and Uranium Corporation 70, 71, 168
Carbon 179
Cathode 104, 168, 169
Cavendish Laboratory 31, 104

CERN (European Organization for Nuclear Studies) 94, 95, 110, 118, 122, 123, 135, 138, 139, 140, 141, 142, 143, 188
Chihuahua 58, 167
Citizenship 53, 71, 72, 76, 91, 93, 168
Cloud chamber (see also Wilson chamber) 77, 150, 158, 161, 172, 173
Cockcroft generator 76
Columbia University 71, 72, 73, 74, 75, 78, 85, 88, 111, 121, 122, 127, 130, 137, 172, 173, 174, 175, 178
Comisión Impulsora y Coordinadora de la Investigación Científica (CICIC) 58, 164, 166
Communists 63, 64, 76, 80
Conservatives 22, 43, 80
Copenhagen 50, 118, 138
Coral Gables, Florida 89, 124, 132, 133, 181
Cosmotron 73, 84, 85, 87, 173, 180, 181
Cyclotron 51, 73, 74, 75, 85, 130, 131, 173, 175, 178, 179

Decay constant 25, 26
Desensitization 35, 39
Deuterium 134, 181
Deutschkreutz 14, 202
Disintegration process (see also Spallation) 46, 50, 149, 159, 192
Disintegration star 5, 10, 44, 45, 46, 47, 81, 91, 94, 159, 160, 162

Dokumentationsarchiv des österreichischen Widerstands 7, 199
Dyes 35, 154

Eastman-Kodak 77, 79, 80
Electrometer 30, 103, 167, 168
Electron 13, 37, 73, 80, 148, 158, 169, 172, 176, 178
Elementary particle 13, 73, 94, 95, 110, 119, 176, 198, 201
Emigrants 81, 93, 98, 129, 197
Emigration 73, 98, 118, 142, 162, 197
Emulsion layer 44, 148, 152, 159, 180
Emulsion stack 180, 181
Emulsion sandwich 178
Escuela Superior de Ingeniería Mecánica y Eléctrica (ESIME) 56, 57, 60, 118, 163, 203
Evanston, Illinois 7, 130
Exile 54, 56, 63, 64
Exposure 34, 39, 135, 141, 145, 154, 159, 183

Fellowship 31, 36, 107, 128, 137
Ferric oxide 170, 171, 193
Florida 7, 88, 89, 90, 91, 92, 98, 99, 110, 118, 132, 136, 137
Florida State University 99, 118, 137
Fordham University 107
Frankfurt 21, 59, 112, 151, 196
Fürstenau, Eppens & Co. (x-ray tube factory) 21

Galvanometer 167, 169
Gamma-radiation (γ-radiation) 34, 146, 152, 154, 165
Gamma-rays (γ-rays) 26, 146, 165, 184, 196
Geiger counter 32, 41, 57, 129
Geneva 54, 94, 97, 110, 118, 139, 141, 142, 143
Gestapo 14
Gibbs Manufacturing and Research Corporation 70
Göteborg 26, 50, 106, 111, 123
Göttingen 16, 36, 37, 38, 39, 40, 42, 115, 116, 122, 141, 142, 191
Grain density 46, 154, 156, 157, 173, 174, 194
Grain density-energy relationship 156, 174
Graz 104, 107, 116, 126

Habilitation 52, 53, 69, 105, 107, 114, 220
Hafelekar 45, 81, 107, 117, 159, 160, 183, 202
Haitinger Prize 31, 104, 105
Half-life 26, 108
Halftone emulsion 159
Hamburg 55, 117
Heinrich Heine Club 63, 64
Helium 62, 105, 134, 158, 165, 181, 186, 199
High-energy particle 44, 73, 78, 173, 178
High-energy physics 100, 102, 139, 173, 175, 184
H-ray (see also proton) 10, 148, 150, 151, 152, 154

Hydrogen 27, 35, 76, 134, 146, 148, 151, 153, 155, 156, 158, 181, 185, 200
Hyperfragment 84, 135, 142, 180, 181, 182, 183, 195, 196
Hyperon 84, 135, 136, 158, 175, 176, 180, 181, 182, 183, 195, 196

Ilford Ltd. 44, 77, 79, 80, 159
Imperial College 47, 51
Innsbruck 6, 45, 107, 117, 142, 159
Institut für Isotopenforschung und Kernphysik 11, 220
Institute for the Physical Bases of Medicine 21, 151
International Federation of University Women 30, 31, 41
International Radium Standard Commission 103
International Rare Metals Refinery 70, 168
Ionization 30, 35, 62, 79, 80, 107, 134, 146, 158, 167, 171, 182, 184, 193
Ionization chamber 37, 45, 68, 158, 171
Ionization parameter 90, 137, 184, 195
Ionization track 35, 44, 131

Janesville, Wisconsin 70, 71, 121
Joachimsthal 18
Jungfraujoch 160

Kaiser-Wilhelm-Gesellschaft 46, 121

Kaiser-Wilhelm-Institut 69, 117, 118, 120
Karlsruhe 30, 115
K-meson (Kaon) 90, 135, 136, 175, 180, 182, 183, 195, 196
Kodak 79, 80

Lainz Hospital 7, 99, 100, 101, 126
Latent image 28, 127, 145, 154
Lead 32, 39, 176, 187
League of Nations 54
Leibniz Medal 91
Leipzig 36, 66, 114, 115, 186, 190, 200, 201
Leopoldstadt 14, 72
Lieben Foundation 44
Lieben Prize 44, 103, 104, 123
London 6, 31, 42, 51, 54, 55, 79, 87, 106, 112, 114, 117, 119, 191, 192, 195, 197, 198, 200, 201, 202, 203
Lugano 100, 124
Luminescence 32, 103, 104, 145, 146, 152, 171
Lung cancer 33, 101

Manhattan Project 111, 115, 116, 121, 173
Marseille 63
Massachusetts Institute of Technology (MIT) 66, 75, 114, 121, 122, 174, 203
Max-Planck-Gesellschaft (Max Planck Society) 7, 117
McCarthy era 76
Meson (see also K-meson, μ-meson, π-meson) 13, 73, 74, 84, 87, 135, 143, 158, 172,

Meson (contd.) 173, 175, 176, 179, 180, 188, 188, 194
Meson production 84, 87, 176, 179, 180, 194, 195
Mexico City 7, 53, 54, 55, 56, 57, 58, 60, 61, 62, 66, 113, 118, 119, 127, 163
Miami 6, 7, 84, 88, 89, 90, 91, 92, 110, 124, 132, 133, 134, 135, 136, 142, 181, 182, 184
Microscope 27, 38, 43, 45, 56, 81, 134, 139, 149, 177, 181, 184
Morelia 61, 62
Morzinplatz 14, 112
Muon, see µ-meson
Music 20, 53, 63, 64, 89, 93, 103, 133

Nature 45, 48, 78, 123, 155, 186, 191, 192, 197
Naturwissenschaften 45, 187, 198
Nazi party 80, 83, 107
Nazis 14, 24, 31, 47, 48, 52, 57, 81, 83, 84, 103, 107, 112, 160, 197
Neutrino 13, 172
Neutron detection 35, 155, 156, 161
Neutron source (see also Polonium-beryllium neutron source) 35, 68, 155
Nevis Cyclotron 73, 130, 131, 173, 175, 178, 179
New York City 6, 70, 72, 73, 74, 111, 119, 121, 173
Nitrogen 27, 148, 149, 179, 186

Nobel Prize 5, 10, 21, 44, 76, 77, 78, 87, 107, 110, 111, 115, 117, 121, 122, 123, 135, 158
NSDAP (see also Nazi Party) 43, 50, 68, 69, 80, 82, 83, 84, 110
NS Teachers' Association 69, 82
Nuclear disintegration 38, 39, 69, 73, 150, 152, 154, 155, 160, 161, 180, 203
Nuclear fission 10, 26, 68, 73, 109, 110
Nucleon 132, 155, 172, 175, 176, 180, 181, 182, 183, 194, 196

Orsay 65, 128
Oslo 48, 49, 50, 51, 53, 54, 55, 107, 108, 117, 118, 127, 160, 162, 163, 164
Österreichische Zentralbibliothek für Physik 7, 49, 53, 77, 112, 114, 115, 118, 121, 124, 125, 126, 185, 187

Paraffin 148, 151, 152, 155, 156, 189
Paris Radium Institute 23, 30, 31, 40, 100, 107, 129, 156
Particle energy 73, 147, 159
Particle identification 90, 153, 174, 182
Patchogue, New York 86, 124
Photographic Society of Vienna 35
Photomultiplier 137, 168, 169
Pinacryptol yellow 39, 115, 154, 156, 159, 186, 203

Index of Subjects   217

Pion, see π-meson
Planck constant 13, 121
Plate group 95, 96, 138, 140
Polonium 25, 26, 27, 30, 31,
   103, 151, 152, 153, 155, 156,
   168, 170, 171, 190
Polonium-beryllium neutron
   source 155, 156
Polytechnical Institute Mexico
   City (Instituto Politécnico
   Nacional, IPN; see also
   Technical University Mexico
   City) 56, 57, 129
Positron 38, 107, 158
Pre-treatment 153, 154, 156
Prong 46, 159
Protactinium 171, 193
Proton
   fast 28, 149, 150, 151, 152,
      154, 157
   recoil 76, 148, 156, 160, 162
Proton energy 153, 156, 157,
   159, 173
Proton track 33, 152, 153, 156,
   157, 159, 160
Pugwash movement 109, 110
Pupin Physics Laboratory 74,
   78

Quark 176, 182

RaA (^{218}Po) 25, 170, 189
RaB (^{214}Pb) 170
RaC (^{214}Bi) 170
Radiation hazards 32
Radiation protection 32, 33
Radioactive decay 26, 28, 148,
   165
Radioelements 32, 169

Radium 18, 21, 25, 32, 33, 62,
   103, 106, 107, 108, 162, 165,
   167, 168, 170, 171, 192
Radon 170, 171
Rahlgasse 7, 16, 35, 113
Range (*Reichweite*) 27, 41, 46,
   147, 149, 152, 153, 159, 162,
   163, 184
Range-energy relation 134, 153,
   156, 184
Rensselaer Polytechnical
   Institute 118, 137

Samarium 52, 162, 163, 193
Saratoga Springs, New York 170
Scanning 94, 130, 131, 134,
   135, 143, 178
Scanning machine (semi-
   automatic device) 75, 131,
   137, 176, 177, 194
Schrödinger Prize 93, 94, 102,
   105
Science Prize of the City of
   Vienna 100
Scintillation 28, 30, 147, 148,
   149, 150, 168
Scintillation counter 136, 168,
   169, 173
Scintillation method 27, 28, 29,
   105, 150, 168
Silver bromide 28, 77, 153, 178
Social democrats 22, 43, 63, 80
Solar radiation 58, 62, 164
Solid-state physics 90, 132
Spallation 159, 160, 162
Spallation star (see also Disin-
   tegration star) 157, 161
Spanish Civil War 63
Spark chamber 134, 136, 185

State Department, U.S.A. 91
Stockholm 7, 118, 124
Streamer chamber 134, 185
Student organization
　Catholic 24, 25
　German nationalistic 24

Tallahassee, Florida 99, 118, 119, 137
Tampa, Florida 7, 91, 136
Technical University Mexico City (see also Polytechnical Institute Mexico City) 49, 53, 54
Track length 152, 153, 156, 159, 181
Trotskyites 63

University of Frankfurt 21, 151
University of Innsbruck 44, 45, 107
University of Miami 88, 89, 90, 110, 132, 133, 136, 181, 184
University of Oslo 48, 49, 107, 108, 162
University of South Florida 91, 136
University of Vienna
　First Physics Institute 32, 106, 109, 120
　Second Physics Institute 17, 18, 21, 34, 68, 69, 80, 104, 106, 109
　Third Physics Institute 69, 106
　Institute for Theoretical Physics 95, 103, 108, 109, 122, 140
Upton, New York 7, 130

Uraninite 167
Uranium 18, 58, 68, 70, 71, 108, 109, 146, 162, 163, 165, 167

Vienna City Council 22, 101
Vienna Jewish Community 7, 113

Wilson chamber (see also Cloud chamber) 41, 77, 150, 158, 161, 162
Women's education 15, 16, 33
World War I 17, 20, 24, 28, 106, 108, 109, 120, 148
World War II 20, 31, 54, 80, 111, 117, 133, 163, 172, 173, 183

X-ray 20, 21, 28, 31, 36, 101, 145, 146, 151

Zelem 14
Zell am See 83
Zinc sulfide (ZnS) 27, 30, 147, 148, 149, 168, 170
Zurich 92, 115, 117, 122

Λ-hyperon 136, 175, 181, 183
μ-meson (muon) 13, 125, 172, 173, 174, 198
Ξ-hyperon 175
π-meson (pion) 5, 10, 13, 77, 79, 84, 87. 90, 135, 136, 172, 173, 174, 175, 178, 179, 180, 181, 182, 195, 196
Σ-hyperon 136, 175, 183
τ-meson 175
Ω-particle 95

INDEX OF PUBLICATIONS BY MARIETTA BLAU

B1 18, 189
B2 21, 151, 189
B3 21, 151, 189
B4 25, 189
B5 33, 151, 152, 189
B6 33, 150, 151, 152, 189
B7 189
B8 33, 153, 189
B9 33, 153, 189
B10 33, 190
B11 190
B12 33, 190
B13 33, 190
B14 190
B15 33, 190
B16 33, 154, 190
B17 33, 153, 190
B18 33, 40, 190
B19 33, 40, 154, 156, 190
B20 36, 191
B21 153, 191
B22 40, 156, 191
B23 154, 156, 191
B24 156, 157, 191
B25 155, 191
B26 146, 147, 148, 149, 191
B27 155, 191

B28 191
B29 191
B30 192
B31 192
B32 192
B33 192
B34 157, 192
B35 159, 192
B36 48, 162, 192
B37 192
B38 76, 159, 160, 192
B39 76, 192
B40 159, 193
B41 52, 162, 163, 193
B42 62, 163, 193
B43 62, 165, 193
B44 62, 164, 193
B45 62, 166, 193
B46 62, 166, 193
B47 62, 167, 193
B48 62, 166, 193
B49 168, 169, 193
B50 169, 170, 193
B51 169, 170, 193
B52 169, 171, 193
B53 169, 171, 193
B54 169, 171, 193
B55 169, 172, 194
B56 174, 194

B57 174, 182, 194
B58 174, 194
B59 175, 194
B60 176, 194
B61 81, 183, 194
B62 131, 176, 194
B63 176, 194
B64 75, 135, 184, 194
B65 131, 176, 177, 178, 182, 194
B66 194
B67 85, 178, 194
B68 84, 179, 194
B69 84, 179, 180, 195
B70 84, 179, 195
B71 84, 179, 195
B72 180, 195
B73 84, 180, 195
B74 184, 195
B75 90, 181, 195
B76 90, 137, 182, 184, 195
B77 90, 195
B78 90, 182, 195
B79 79, 92, 96, 135, 184, 195
B80 90, 135, 182, 196
B81 96, 184, 196

AUTHORS

Robert Rosner, born in Vienna in 1924, emigrated to England on a children's transport in 1939, worked during the war in an engineering company while continuing his education in evening classes (general certificate of education in 1942), returned to Austria in 1946, studied chemistry at the University of Vienna 1947–1955, Ph.D., worked 1956–1990 in a chemical company as a chemist and finally as sales manager. Studied political science and history of science at the University of Vienna 1991–1997, Mag.Ph. Present address: Paul-Heyse-Gasse 32, A-1110 Vienna; Robert.Rosner@tele2.at.

Brigitte Strohmaier, born in Vienna in 1948, studied physics at the University of Vienna 1967–1974, Ph.D. *sub auspiciis praesidentis*, received certification to teach nuclear science at the university (*Habilitation*) in 1988 and worked as a university lecturer (*tit. ao. Prof.*) at the Institut für Isotopenforschung und Kernphysik der Universität Wien (formerly Institut für Radiumforschung und Kernphysik der Österreichischen Akademie der Wissenschaften). Early retirement in December 2003. Present address: Währingerstraße 17, A-1090 Vienna; Brigitte.Strohmaier @univie.ac.at.